Praise for

Did You Just Eat That?

"Myth busters move over and squeamish eaters beware. Here is a fun, entertaining, fact-filled and scientific dive into the potentially daily disgusting and dangerous encounters we can have with food—but also with easy-to-follow advice to save you!"

—Rachel Herz, author of *Why You Eat What You Eat*

"If, like me, you've always been suspicious of hand driers in toilets, or worried about the transfer of germs when you pick up a menu in a restaurant, then this book is for you. Entertaining and enlightening in equal measure."

—Mark Miodownik, author of *Stuff Matters*

"While fascinating and often hilarious, the experiments in *Did You Just Eat That?* could make you live in fear of double-dippers, hand blowers, passed popcorn, and the things you put in your Bloody Mary. Here's to more good microbes than bad, and to Paul Dawson and Brian Sheldon's wanting you to meet their little friends!"

—Michael Feldman, host of *Whad'Ya Know?*

Did You Just Eat That?

Did You Just Eat That?

**TWO SCIENTISTS EXPLORE DOUBLE-
DIPPING, THE FIVE-SECOND RULE,
AND OTHER FOOD MYTHS IN THE LAB**

Paul Dawson and Brian Sheldon

W. W. NORTON & COMPANY
INDEPENDENT PUBLISHERS SINCE 1923
NEW YORK   LONDON

For information about permission to reproduce selections from this book,
write to Permissions, W. W. Norton & Company, Inc., 500 Fifth Avenue,
New York, NY 10110

For information about special discounts for bulk purchases, please contact
W. W. Norton Special Sales at specialsales@wwnorton.com or 800-233-4830

Manufacturing by Versa Press
Book design by Daniel Lagin
Production manager: Lauren Abbate

Library of Congress Cataloging-in-Publication Data

Names: Dawson, Paul L., 1954– author. | Sheldon, Brian W., 1950– author.
Title: Did you just eat that? : two scientists explore double-dipping, the five-second
    rule, and other food myths in the lab / Paul Dawson and Brian Sheldon.
Description: First edition. | New York : W. W. Norton & Company, [2019] |
    Includes bibliographical references and index.
Identifiers: LCCN 2018030230 | ISBN 9780393609752 (hardcover)
Subjects: LCSH: Food—Microbiology.
Classification: LCC QR115 .D28 2019 | DDC 579/.16—dc23
LC record available at https://lccn.loc.gov/2018030230

W. W. Norton & Company, Inc., 500 Fifth Avenue, New York, N.Y. 10110
www.wwnorton.com

W. W. Norton & Company Ltd., 15 Carlisle Street, London W1D 3BS

1 2 3 4 5 6 7 8 9 0

We'd like to dedicate this book to all germophobes and hope they live life to the fullest.

Contents

Prologue

Why write a book about the five-second rule, double-dipping, and other food myths? Well, because inquiring minds want to know. The truth is, we've always been puzzled and a little impressed by the faith people seem to have in catchphrases for a host of food-handling behaviors. Five seconds? Says who? And double-dipping. George Costanza from *Seinfeld* tends to be wrong about things. So, could Timmy, the character who calls George out for this practice in a famous episode, be right? The more we thought about it, the more examples of questionable food-handling behavior came to mind. Some of them don't even have catchphrases yet. Whether it's the communal popcorn tub at the movies or the Ping-Pong ball used in the game of beer pong, most of us don't think twice about what else besides popcorn or beer we might be putting in our mouths.

Catchphrases will come in time, but we wanted to take a closer look at what's really going on. That's why, together with our under-graduate students from the Creative Inquiry Program at Clemson University, we conducted bona fide scientific research in a faculty-mentored team environment. This was our way of introducing the scientific method to students who at first may not be interested in research and getting them to look at common, everyday topics and "beliefs" about bacteria and food. The questions raised captured our participants' interest in their own and their classmates' behaviors. What we found amazed us, and we decided to share our findings with a wider audience—including you, the reader.

After a series of research experiments, we hatched the idea to com-pile a book on these topics for the everyday reader. In each chapter, we present results from our original published research, along with research by others, on how bacteria are transferred and survive on foods and con-tact surfaces we are exposed to in everyday life. We hope this book will both enlighten and entertain you.

So, how should you enjoy this book?

After the introduction, which gives you some basic information about the mysterious microbial world, the chapters are grouped into three parts covering how bacteria spread on (1) different surfaces, (2) air and water, and (3) food transport mechanisms. In each chapter, we first share some trivia and background on that particular topic, then describe do-it-yourself experiments we have conducted to answer the question posed. Near the end of each chapter, you'll find other research completed on the same topic to give you a better grasp of the myth or issue being considered. Those who want to know the details or try the experiments themselves, including both materials and methods, can read the sections titled "Science Stuff Ahead." Readers who are more interested in the results or in discussions contrasting other research can skip those sections. At the end of each chapter, you'll find a section called "Things to Consider." In it we look at serious implications for the findings as well as some humorous subjects, such as the total amount of dog feces (20 billion pounds in the United States per year) on the ground where you might drop food—fun and relevant stuff like that.

Things to Consider

In this book, you will see references to statistics in each chapter, such as "statistical significance" and "significance at the 5 percent level." Statistics is the science concerned with the collection, organization, and analysis of information. To make predictions about populations, we typically use information from a sample taken from the population. Since the sample contains only a small fraction of the population, predictions about the population can sometimes be wrong. When results are said to be "statistically significant" or "statistically different" this implies that the results based on the samples are unlikely to happen by mere chance—the treatment had an effect on results. Being "significant at the 5 percent level" or "the significance level is at 5 percent" means there is, at most, a 5 percent chance of concluding there is a statistical difference when a difference does not really exist.

Although all studies in this book are referred to as "experiments," the authors do acknowledge that some of the studies are in fact observational studies. For consistency purposes, we have labeled all studies as "experiments."

Feel free to jump to a chapter that might interest you and read any part. Then go back and read about the details later.

Did
You
Just
Eat
That?

A DIVE INTO THE MYSTERIOUS MICROBIAL WORLD

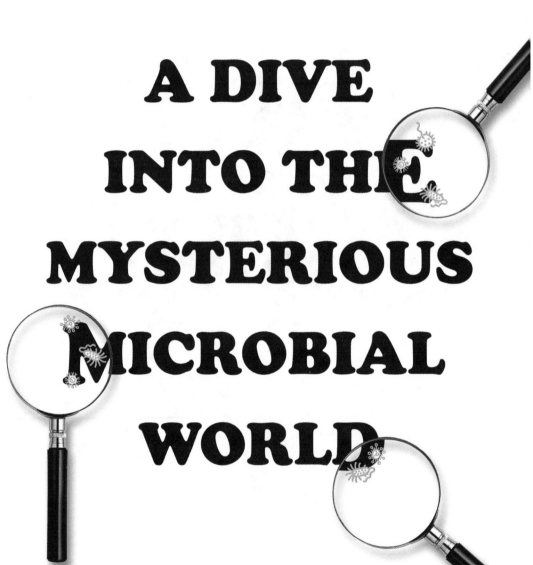

Introduction

A DIVE INTO THE MYSTERIOUS MICROBIAL WORLD

People live in a world where bigger is better, but this book explores the importance of some of the smallest organisms on earth. As they say in the happiest place on earth, it's a small world, after all. We can't see these organisms, but they are there. In fact, there most likely isn't any place you *won't* find them. Even as you read this page they are all over you, and in you, and covering everything you can see or touch. These organisms make up the world of microbes—bacteria, viruses, fungi, and other invisible living organisms. They're called microorganisms because, as individual cells, they are invisible to the naked eye and can be seen only through a microscope.

As the writers of this book, we feel that the better you understand the microbial world, the more you will profit from and enjoy the book. We start with a brief overview of the microbial world to give you some of the basics you'll need to follow the discussions and experiments presented throughout. In the chapters that follow, we focus mainly on examining and testing several common food myths, mostly about bacteria. The myths we tackle involve games with food, food-handling practices, and human behaviors that might end up transferring microorganisms from one surface to another, and potentially on to you.

Many of you probably have seen the hilarious Christmas movie classic *National Lampoon's Christmas Vacation* with Chevy Chase (aka Clark Griswold). Do you remember the scene where the extended family is seated around the dining room table, expectantly waiting to eat Christmas dinner? Remember when Cousin Eddie—yes, the man with a bigger heart than brain—takes a spoonful of sweet potato casserole using the serving spoon, tastes the food, fills his plate, and *then* returns the serving spoon back to the casserole dish? Although Cousin Eddie's eating antics were hilarious, his table etiquette leaves a lot to be desired. He could greatly benefit from a refresher course on Dining Etiquette 101 as it pertains to safe food-handling practices. We've included this example just to demonstrate that human behavior is fickle and not easily controlled. We bet you have observed similar questionable behaviors in people of all ages.

People sometimes have odd eating behaviors. Ever heard of xylophagia (eating wood), trichophagia (eating hair), urophagia (drinking urine), or even coprophagia (eating feces)? Back in 2009, astronauts actually drank urine that had been "peecycled" through a $250 million wastewater recycling system on the International Space Station. Depending on the microorganisms that might be present on a surface—for example, on a serving spoon or Ping-Pong ball (which we'll discuss in Chapter 2)—the consequences of these behaviors could lead to microbial infections or disease for those engaged in such activities.

Microorganisms make up a huge and incredibly diverse fraction of all living organisms that occupy every habitat on this earth: air, soil, rock, deep oceans, boiling-hot springs, arctic snow, plants, animals, humans, food, food contact surfaces, and more. Their diverse habitat illustrates how extremely adaptable microorganisms are to nearly every conceivable environment and condition. Scientists have estimated that the number of bacteria on earth is approximately 5×10^{30}—that's the number five followed by thirty zeros! The total biomass of these bacteria exceeds that of all plants and animals.[1] More recent estimates have lowered that number to somewhere between 9.2×10^{29} and 31.7×10^{29}, which is still an unbelievably large number.[2] To look at it another way, the number of bacterial cells on and in your body is estimated at ten times more than the number of all the cells contained in your body. (In total, these bacteria weigh about 4 pounds, and most of them reside in your gastrointestinal tract and on your skin. So when your better half hints that you could stand to lose a few pounds, you can blame some of that extra weight on all that pesky bacteria hitchhiking on your body.) Equally impressive, the number of bacteria currently residing in and on one person is greater than the total number of people who have ever lived or are currently living on our

ALL HAIL THE MIGHTY BACTERIA!

planet. Ironically, humans in theory have evolved from single-celled bacteria starting with a process called endosymbiosis. Based on one study, the guesstimated total human population has been set at 108.2 billion. That's 1.08×10^{11} humans, a far cry from the 5×10^{30} total bacterial estimate.[3]

Many microorganisms have very little or seemingly no direct impact on us, while others can directly affect us in either desirable or undesirable ways. Those aiding in the digestive processes in our gastrointestinal tract or in the production of fermented foods (including yogurt, cheeses, and meats) are very desirable. But those microorganisms causing food spoilage and foodborne diseases that can lead to significant personal discomfort and even death are, well, extremely undesirable.

Next we'll look briefly at the diversity, size, shape, and population of different microorganisms, especially the growth and reproduction of bacteria and viruses, followed by a short discussion of diseases associated with foods. Do not, however, be alarmed by all of this. By far, most microorganisms benefit humans by eliminating waste, helping plants grow, and creating tasty foods. Also reassuring is the estimate that less than 1 percent of the known bacteria will actually make you sick. After all, most of us don't generally get sick from eating even after consuming three meals a day, 365 days a year. But that doesn't mean you should play the odds and blatantly ignore using your common sense and good food-handling practices. If you're getting at all squeamish, try to relax. We promise to share the best ways to prevent, reduce, or eliminate bacteria from your food and any food-related environments. At the end of the book, you'll find some good food-handling and preparation practices as well as effective cleaning and sanitation procedures.

DIVERSE MICROORGANISMS

The cells that make up plants, animals, and humans exist only as part of multicellular structures. Bacteria, on the other hand, are able to live as single-cell microorganisms in nature. They can carry out all the necessary life functions such as growth, energy metabolism, reproduction, communication, and movement. Microorganisms have existed since the beginning of time, and they've had billions of years to evolve and adapt to nearly every environment. Therefore, biologists see immense diversity of microorganisms in nature. They also comprise the largest population and biomass of any living thing. While many bacteria have already been identified, many more are yet to be discovered and characterized, and presumably only about half of the bacterial species can be cultured in the laboratory today. For this reason, it is impossible to truly estimate the enormity of their population and total impact they have had on this earth and on people.[4]

As you will see in the next section, microorganisms come in various shapes and sizes, even though they are all very small. Thanks to their genetic diversity and ability to acquire energy from nearly every conceivable source, including light, different species of bacteria have been discovered in nearly every habitat and extreme environment on earth.

These hardy little bugs can even be found in many unlikely places such as boiling-hot springs (235°F); ice covering lakes, glaciers, and polar seas (32°F); salty bodies of water (15 to 32 percent salt); harsh acidic and alkaline water and soil environments (pH of 0 and 12, respectively); and even environments lacking oxygen (for example, the anaerobic bacteria *Clostridium*).[5]

SIZE, SHAPE, AND POPULATIONS

Scientists call the study of these very small organisms microbiology. In the late 1600s, the Dutch scientist Antonie van Leeuwenhoek became the first person to catch a glimpse of these tiny creatures.[6] Being a draper by trade (he was a dealer of cloth), he was greatly interested in each thread of the material. He wanted to examine an individual thread up close, so Leeuwenhoek constructed the first crude microscope. After several refinements, he had a device that could magnify an object some 275 times.[7] His most powerful microscope now resides at the Utrecht Museum in the Netherlands. As Leeuwenhoek perfected his gadget, a whole new world was revealed to him. These once imperceptible "beings" were suddenly out in the open. He was amazed and captivated by discovering these invisible creatures as well as many other treasures revealed by the microscope.

Leeuwenhoek reported seeing "a great multitude of living creatures in one drop of water amounting to no less than 8 or 10 thousand, and they appear to my eye through the microscope as common sand does to the naked eye." He called these microscopic living creatures animalcules, and in 1683 he reported seeing different types of animalcules in spittle and tooth scrapings. Have you ever wondered about that morning breath of yours and considered why just after waking up was not a good time for that morning kiss? Well, you can blame much of that halitosis breath on germs—the animalcules that grew on those pearly whites of yours while you were sleeping. Now, *animalcules* is a word you don't hear every day; it sounds more like a new toy or type of animal cracker. *Spittle* has come to describe those tiny droplets of saliva that become airborne during conversation; in a sentence, it is used like this: "I was talking to my boss and spittle flew out of my mouth and landed on her nose." Leeuwenhoek's 1683 letter to the Royal Society of London contained drawings of bacteria showing their common shapes including bacilli (rod-shaped), cocci (circular), and spirillum (spiral-shaped).[8]

Microbes such as bacteria and viruses are considerably smaller than plant, animal, and human cells. Bacteria are typically around ten times smaller than a human cell; most viruses are a hundred times smaller. So, for instance, if a bacterial cell were the size of a cat or small dog, then a human cell would be the size of an adult human. Using the same scale, viruses would be about the size of a mouse.

Biologists generally measure the size of microorganisms in units of micrometers (also called microns, abbreviated as μm), which are one millionth of a meter, or 0.00004 inch. To give you some idea of this scale, the period at the end of this sentence is around 0.5 millimeters (mm) or 500 μm. An average-sized, round-shaped bacteria (about 1 μm in diameter) is about five hundred times smaller than a period, meaning it would take about 500 bacteria to fill one period. Bacterial cells can range from around 0.15 μm to over 700 μm in diameter, so the largest cells are nearly visible with the naked eye.[9] The diameter of viruses ranges from 0.02 to 0.3 μm.

Even though microorganisms are incredibly small, they make up very large populations in nature. For example, 1 gram of soil typically contains 40 million bacterial cells, and 1 milliliter (mL) of fresh water (say, from a river or lake) can contain 1 million bacterial cells.[10] To illustrate the full scope of these numbers, suppose we took the estimated population of bacteria on this earth (5×10^{30}) times the average size of a bacterium (around 1.5 μm) and then divided this product by the average distance from the earth to the moon in micrometers (238,900 miles, or 3.84×10^{14} μm). If we then lined up the total bacteria population side by side, they would account for 9.76×10^{15} round trips to and from the moon. Now that, dear reader, is a lot of bacteria!

Besides their small size, bacterial cells come in several shapes, including a sphere (termed a coccus; plural is cocci), rod or cylinder

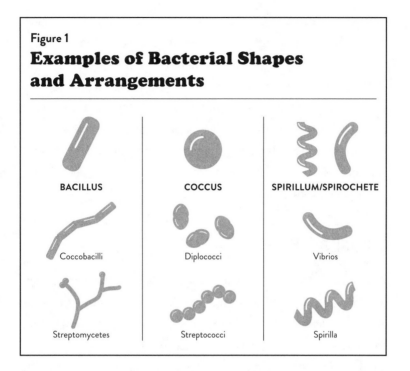

Figure 1

Examples of Bacterial Shapes and Arrangements

BACILLUS	COCCUS	SPIRILLUM/SPIROCHETE
Coccobacilli	Diplococci	Vibrios
Streptomycetes	Streptococci	Spirilla

(termed bacillus or bacilli), curved rod, and spiral (Figure 1). These organisms can exist as single cells, or they may be grouped together to form pairs of cells (we call them diplococci), short or long chains (such as *Streptococcus*), irregular or grapelike clusters (*Staphylococcus*), filamentous bacteria (long, thin cells or chains of cells), or budding and appendage-shaped cells.[11] Most of these bacterial cells exist in nature as actively growing and reproducing organisms when conditions are suitable. However, when the conditions become unsuitable—such as temperature extremes, drying conditions, or lack of nutrients—some species of bacteria like *Clostridium* and *Bacillus* can produce highly resistant dormant spores through a survival process we call sporulation. These specialized cells are considerably more resistant than actively growing cells to heat, harsh chemicals, disinfectants, and even radiation. Unfortunately for us, this sporulation process of some bacteria makes it more challenging to kill them with typical surface disinfectants like those used in hospitals, restaurants, and food processing plants. When conditions become more favorable for growth, these spores will germinate to an actively growing vegetative state. Then they can cause food spoilage and shortened product shelf life as well as potential foodborne illness problems for the food industry and consumer.

VIRUSES

We're sure that anyone reading this chapter has had the common cold or flu, or has heard of polio, smallpox, Ebola, and HIV/AIDS. All these and many more diseases are caused by viruses. Even though scientists and doctors have made great strides in eradicating some of these diseases through vaccination programs and other treatments, Ebola and AIDS (acquired immunodeficiency syndrome) and other viral diseases still cause many deaths around the world. Examples of two worldwide pandemic diseases involving viruses are the 1918 Spanish flu pandemic and the HIV (human immunodeficiency virus) pandemic. The Spanish flu killed over 100 million people, or about 5 percent of the world's population in 1918. Today an estimated 38.6 million people worldwide have the HIV virus living in them.[12]

More relevant to this book are foodborne diseases caused by viruses, although when it comes to some human behaviors and practices, we cannot rule out the transmission of other viruses not generally associated with foods. The most common foodborne illness or gastroenteritis (a diarrhea and vomiting illness) is caused by the norovirus, commonly known as the stomach flu or the winter vomiting bug. This virus is highly contagious and can tear through cruise ships, classrooms, and other crowded spaces such as day care centers. We'll bet that all of you have heard about those "cruises from hell" where hundreds of vacationers contracted the norovirus while aboard the ship. We get it—sitting in the bathroom is not the most desirable place to spend your vacation. According to estimates from the Centers for Disease Control and Prevention (CDC),[13] the norovirus causes around 19 to 21 million cases of acute gastroenteritis annually in the United States, or over half of all foodborne disease outbreaks.

Viruses are small infectious agents that replicate, or reproduce, only within living cells of other organisms. You might call them parasites, since they cannot reproduce on their own but require a host cell. The average virus is about one-hundredth the size of the average bacterium, or between 0.02 and 0.3 µm. Viruses are capable of infecting all living organisms: plants, animals, and bacteria. Millions of viruses are found in nature, but only around five thousand have been characterized in detail.[14] Like bacteria, viruses are found in nearly every ecosystem on earth, and they are the most abundant type of living organism.[15] They are able to exist, but not reproduce, outside the host organism for extended periods of time, so they can be transmitted between hosts. When outside of a host, they exist as independent viral particles called virons. Virons contain DNA or RNA, a protein coat, and sometimes an outside envelope. They come in many shapes, ranging from simple helical and icosahedral forms to much more complex structures, as shown in Figure 2. Notice the shape of the bacteriophage virus, which infects bacteria. It might remind you of a lunar landing craft, or maybe a spider.

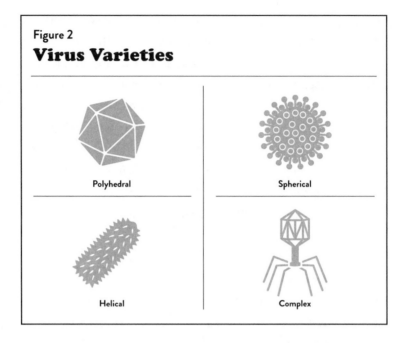

Figure 2
Virus Varieties

Polyhedral	Spherical
Helical	Complex

When it is outside a host cell, the viron is metabolically inactive, like a dormant bacterial spore, but it can be transmitted between host cells. Once it's injected into the host cell—for example, one of your cells—virus replication can begin. This process consists of producing new copies of the virus genome and synthesizing the viral coat components.[16]

Viruses are clever in that while they are in the host, they depend heavily on the host's cell structure and metabolic components. For example, they are capable of redirecting the host's metabolism to support their own replication and production of new virons. Now that's what we call invasion of the body snatchers. Ultimately these new virons are released from the host, and the process continues in another host cell. You'll learn more about viruses later in this chapter when we discuss foodborne illnesses.

WHAT MAKES MICROORGANISMS TICK?

Like people, all microorganisms require nutrients to survive. In more technical terms, they have defined nutritional requirements that they must obtain from their environment to sustain their growth and reproduction. All cells, including our own, require a carbon source—generally an organic one—that can be supplied by many nutrients including amino acids (from proteins), fatty acids (from fats), organic acids, sugars (from carbohydrates), nitrogen bases, and other organic compounds. Some bacteria are autotrophic (able to make their own food) and are even capable of drawing some or all of their carbon requirements from atmospheric carbon dioxide (CO_2). Scientists have taken advantage of these organisms' odd appetite for CO_2 in several interesting applications. For example, Russian scientists have experimented with using the autotrophic bacteria *Ralstonia eutropha* for scrubbing carbon dioxide from their manned spacecraft and to serve as a potential food source for their astronauts. Closer to home, Massachusetts Institute of Technology scientists have been testing a genetically modified form of this same bacteria to make alcohol fuels from carbon dioxide and hydrogen for use in transportation.[17]

Another nutritional requirement of microorganisms is nitrogen, which is critical for synthesizing protein and nucleic acids such as DNA. In nature, nitrogen is generally available as ammonia, nitrate, or nitrogen gas. Most bacteria can use ammonia and some nitrate as their nitrogen source, but only nitrogen-fixing bacteria commonly found in the soil can use nitrogen gas. Without these nitrogen-fixing bacteria that convert nitrogen gas to ammonia or other nitrogen-containing compounds, farm-

You're welcome!

ers could not grow their crops, and life would not exist as we know it.

As an example, to ensure optimum growth of legumes—beans, peas, soybeans, and others—seed companies typically inoculate their seeds with commercial cultures of species of the nitrogen-fixing bacterium *Rhizobium*.[18] Other nutrient requirements are the minerals essential to growth, called macronutrients: phosphorus, sulfur, potassium, magnesium, calcium, sodium (not an essential nutrient for all microorganisms), and iron. Besides these essential macronutrients, many microorganisms require very small amounts of micronutrients (referred to as trace elements) to maintain critical cell functions. A few of these are boron, cobalt, copper, manganese, and zinc. Other growth factors that microorganisms must have are vitamins and amino acids. As you can see, their nutrient requirements are not that different from ours. Many organisms can synthesize some of these compounds. Other organisms can obtain them only from the environment, such as the gastrointestinal tract of animals and humans, plant tissues, and the soil, to name a few.[19]

Besides nutritional requirements, the ability of microorganisms (except viruses) to grow in an environment such as the milk stored in your refrigerator is determined not only by the food itself but also by the environment in which the food is stored. These two conditions are designated as *intrinsic* (inherent nature of the food tissues) and *extrinsic* (properties of the storage environment) *factors*. Intrinsic food factors affecting microbe growth include:

- type of food
- pH (acidity or alkalinity)

- availability of water (referred to as water activity)
- presence or absence of oxygen (oxidation-reduction potential)
- antimicrobials (such as essential oils in spices, lysozyme in eggs and milk)
- physical structures like the covering around nuts, eggs, fruits, and animals (shells, skins, and hides) that prevent entry of microorganisms

Extrinsic factors include the environmental conditions in which the food is stored:

- temperature
- relative humidity
- presence and concentration of inhibitory gases such as carbon dioxide and ozone
- presence and activities of other microorganisms that produce inhibitory substances targeting other microorganisms

If you have too much or too little of these different factors, you will either promote, stop, or retard microbial growth. Although we will discuss some of them separately, it's important to realize that in most environments such as food systems, some or all of these factors are present together. Therefore they will have a cumulative effect—either favorable or unfavorable—on microbial growth.

The first intrinsic factor to consider is pH, the acidity or alkalinity of the environment. Like us, each organism has a pH range within which growth is possible and usually an optimum pH for growth. Similarly, we humans wouldn't survive for long if we consumed a large volume of a strong acid like hydrochloric acid or a strong alkaline solution like lye. Acidity and alkalinity of a food or other environment is expressed on a logarithmic pH scale that extends from 0 to 14 (0 is extremely acidic, 7 is neutral, and 14 is extremely alkaline). Most natural environments have pH values between 5 and 9 and contain the greatest variety of microorganisms. Some bacteria and fungi can grow at pH values of less than 2, such as in lemon juice and gastric juices found in our stomachs, or greater than 9, as found in soda lakes and alkaline soils; these are referred to as acidophiles or alkaliphiles, or acid- and alkaline-loving microorganisms. The pH is an important factor in controlling the growth of microorganisms in acidic foods like dill pickles and in fermented foods including pepperoni, sauerkraut, and yogurt. Because organisms have a pH growth optimum

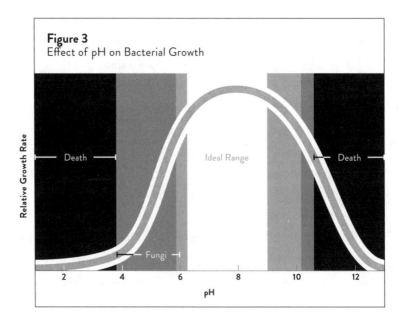

Figure 3
Effect of pH on Bacterial Growth

Relative Growth Rate

Death

Ideal Range

Death

Fungi

pH

2   4   6   8   10   12

and range, when they are subjected to an environment outside their pH range, they will stop growing and eventually die (Figure 3).[20]

A second important intrinsic growth factor is the availability of free "unbound" water required for microbial growth. Free water, in contrast to bound water, is water that is not tied to other substances, such as macromolecules like sugars and cell structures. Free water is therefore available to microorganisms to support their different cellular functions. You might think of free water as water that can be extracted easily from foods by squeezing, cutting, or pressing whereas bound water cannot be easily extracted.

Like us, microorganisms cannot grow without water. Water in foods is measured in terms of water activity (a_w) and is expressed as a value from 0 to 1, where 1 is complete saturation and 0 is complete dryness. Water activity differs from water content since water activity takes into account the ability of certain food components like salt and sugar to "bind" water, preventing bacteria from being able to utilize it. Similar to pH, water activity also has an optimum and range in which various microorganisms can grow. We group foods into the following a_w values:

- 0.10 to 0.20 a_w (cereals, crackers, dried milk, sugar)
- less than 0.60 a_w (noodles, chocolate, dried egg)
- 0.60 to 0.85 a_w (jelly, dried fruits, nuts, Parmesan cheese)
- 0.85 to 0.93 a_w (fermented sausage, maple syrup, dried cured meat)
- 0.93 to 0.98 a_w (evaporated milk, bread, tomato paste, processed cheese)
- 0.98 to 0.99 a_w (fresh meat, fruits, vegetables, milk, eggs)

Molds generally grow at a minimum a_w of 0.8, most kinds of yeast at 0.85, and bacteria at 0.90 to 0.93. However, there are exceptions to these general values. For example, the bacteria *Staphylococcus aureus* (responsible for a foodborne illness and skin infections) can grow at an a_w of 0.85. Humans have been using the food preservation practice of lowering water activity by drying since the beginning of civilization (meats, fish, fruits, vegetables, and milk are just a few examples).

A third critical intrinsic factor is the presence or absence of oxygen, something we refer to as oxidation-reduction potential. Although we tend to think that all life requires oxygen to survive, many microorganisms can and must live in environments either absent of oxygen or in very low concentrations of oxygen. Organisms that can grow and use oxygen are called aerobes, while those that can grow at only low oxygen levels are termed microaerophiles. Facultative microorganisms can grow in the presence or absence of oxygen. Conversely, many organisms cannot grow in the presence of oxygen and are referred to as strict or obligate anaerobes. Aerotolerant anaerobes are organisms that can tolerate oxygen and grow but do not use oxygen.

To control the growth of some microorganisms that cause aerobic spoilage in foods, manufacturers use such processes as canning at high temperature and vacuum packaging to remove air. At times food processors must use other microbial control measures, such as lowering the pH to increase the acidity of the product and thus control the growth of anaerobic bacteria (for example, *Clostridium botulinum*, which causes botulism in some canned food products).

Of the extrinsic factors—the environment where the microorganisms exist—temperature is probably the most important environmental factor controlling microorganism growth and survival. Besides their need

DID YOU JUST EAT THAT?

for favorable pH, a$_w$, and oxidation-reduction potential, microorganisms are affected by minimum, maximum, and optimum growth temperatures. When exposed to conditions outside their temperature range, they will stop growing and may eventually die. At their optimum growth temperature, microorganisms will grow at their maximum rate when other conditions such as nutrient availability are favorable. The growth temperature ranges of microorganisms vary widely, which is one reason they exist everywhere on earth. Some grow in extremely cold (antarctic ice) or hot (hot springs and geysers) environments, and others have optimal temperatures similar to warm-blooded mammals like we humans.

Figure 4 shows microorganisms according to their temperature optimums: they are psychrophiles, which love the cold (around 4°C, or 39°F); mesophiles, which are close to our body temperature (around 39°C, or 102°F); thermophiles, which can take the heat (around 60°C, or 140°F); or hyperthermophiles (around 88°C to 106°C, or 190°F to 223°F).

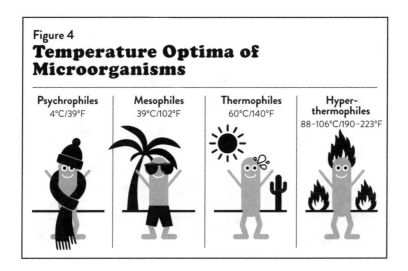

Figure 4

Temperature Optima of Microorganisms

Psychrophiles	Mesophiles	Thermophiles	Hyper-thermophiles
4°C/39°F	39°C/102°F	60°C/140°F	88–106°C/190–223°F

Have you ever wondered why the milk in your refrigerator eventually spoils? Well, you can blame it on those psychrophiles like pseudomonas, which can grow slowly at refrigeration temperatures. Most bacteria that cause foodborne disease fall in the mesophile classification and include *Salmonella, Campylobacter,* and *Staphylococcus.* In the psychrophile group a few other bacteria, such as *Listeria,* can cause foodborne disease. Besides their optimum growth temperature, microorganisms can grow over a wide range of temperatures. For instance, different strains of *E. coli* can have minimum and maximum growth temperatures from 8°C to 48°C (46°F to 118°F), which is an extremely wide range. As with the other intrinsic and extrinsic growth factors, food manufacturers use processing and storage temperatures to prevent the growth of microorganisms present in the food and ultimately destroy those microorganisms. Canning, pasteurization, cooking, refrigeration, and freezing of foods are some thermal processing methods used to eliminate or prevent growth of undesirable microorganisms in foods. It is also critical to hold foods at proper temperatures outside the danger zone of 40°F to 140°F to prevent the

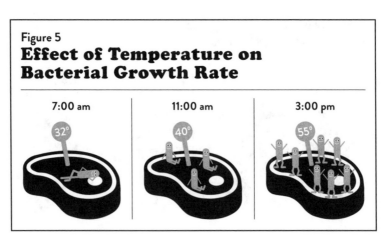

Figure 5
Effect of Temperature on Bacterial Growth Rate

growth of both food spoilage organisms and foodborne pathogens (Figure 5).[21] That's why it is so critical for restaurant workers to keep the temperature of their hot and cold salad bars outside this danger zone—40°F or less for foods that need to be served cold and above 140°F for hot foods.

GROWTH AND REPRODUCTION

For all living things, the name of the game is life over death. Microorganisms are no exception to this basic fact of life. Because the life span of an individual bacterial cell is fairly short, cell growth and reproduction are essential processes for perpetuating the species. Unlike multicellular organisms including plants and animals, in bacteria the increase in cell size and reproduction by cell division are closely connected. When environmental conditions are favorable, most bacteria grow to a fixed

size and then reproduce through a process called binary fission, a form of asexual reproduction. This means they don't actually need a partner.

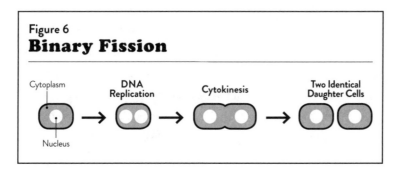

Figure 6
Binary Fission

Cytoplasm   DNA Replication   Cytokinesis   Two Identical Daughter Cells

Nucleus

Binary fission is a reproductive process by which one cell simply divides into two new daughter cells (Figure 6). When this occurs, it is referred to as one generation where the population doubles. Under growth conditions where nutrients and water are available and the temperature, pH, and oxidation-reduction potential are optimal, bacterial cells can double in population in about 15 to 20 minutes, and some cells can do so in less than 10 minutes.[22] Thankfully for us, most organisms grow and divide at slower rates because the growth conditions are not always optimal.

If you were to examine a bacterial population over time, you would observe four distinct phases of the growth cycle. The first phase is when a population of bacteria first enters an environment that may or may not be rich in nutrients. Under these conditions, the cells must start adapting to their new environment. This is called the lag phase, a period of slow growth and a time when the cells are adjusting and preparing for rapid growth. The second phase is the log phase (logarithmic or exponential phase), which is characterized by extremely rapid cell growth and division. In this phase, the cells are at their highest rate of growth and cell division. The time it takes for the population to double during this phase is called the generation time or doubling time. In this phase, microorganisms rapidly use the available nutrients until a nutrient is depleted; when this happens, their growth becomes limited. Theoretically, if a single bacterium having a 20-minute generation time is allowed to grow exponentially for 48 hours or 144 generations, the resulting population would weigh four thousand times more than the earth. This example is mind-boggling and unbelievable considering that the cell weighs only about one-trillionth of a gram.[23] Such an extreme case of bacterial growth is highly improbable since nutrient depletion and buildup of toxic waste products generally limit growth well before bacteria reach these massive populations.

The third phase of the growth cycle is called the stationary phase. It is the result of depleted nutrients or buildup of waste products that inhibit growth. In this phase, cell metabolism slows significantly, thus reducing growth and initiating the transition from rapid growth to a stress response state, such as from the lack of nutrients. In this stage, cells undergo no net increase or decrease in number.

The final phase is the death phase. It takes place when the bacteria run out of nutrients and die. Figure 7 is a typical bacterial population growth curve. Time in hours is recorded on the horizontal axis, and bacterial population is expressed as a logarithmic value on the vertical axis. The shape of the growth curve can be influenced significantly by the history of the organisms (their current physiological state) as well as the influence of the intrinsic and extrinsic growth factors previously

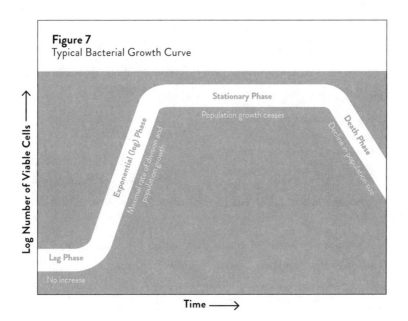

Figure 7
Typical Bacterial Growth Curve

Log Number of Viable Cells ⟶

Stationary Phase
Population growth ceases

Death Phase
Decline in population size

Exponential (log) Phase
Maximal rate of division and population growth

Lag Phase
No increase

Time ⟶

discussed. For instance, when cells are exposed to minimum nutrients, growth temperature, or pH instead of optimal conditions, the lag phase of this curve may be significantly longer because the organism must adjust to its more limited environment. Changes in the duration of the log, stationary, and death phases may also occur due to changes in the environment.

So, practically speaking, why do you need to know how rapidly bacteria double their population? Well, consider the genus *Salmonella*, a bacterial group that in 2014 was responsible for about 7,500 confirmed foodborne illness cases per 100,000 population in the United States. If you exposed these bacteria to ideal growth conditions such as those present in a chicken casserole allowed to sit unrefrigerated for 4 hours at 90°F on a summer day, you could determine what the cell population would be after 4 hours. (Sounds like a family reunion picnic, doesn't it?)[24] With an initial cell population of 40 *Salmonella* cells per gram of product and a cell generation time of 15 minutes, after 4 hours the population would have grown to 2,621,440 cells per gram of product (Figure 8).

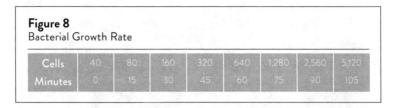

Figure 8
Bacterial Growth Rate

Cells	40	80	160	320	640	1,280	2,560	5,120
Minutes	0	15	30	45	60	75	90	105

Imagine what the final population would be if the starting population had been a thousand or more cells per gram of product. Try 65,536,000 cells per gram. It's something to ponder when food has been left at room temperature for too long.

FOODBORNE DISEASE

Do you know that people living in the United States have about a one in six chance (about 17 percent) of coming down with a foodborne illness this year? That's what the Centers for Disease Control and Prevention (CDC) estimate.[25] Of these cases, they estimate that 128,000 Americans just like us will go to the hospital, and 3,000 of us will die. Not very encouraging news, is it? This discussion of foodborne illness, particularly illnesses associated with bacteria, will be helpful if you consider engaging in some of the food-handling practices described in the following chapters. In the United States, foodborne diseases caused by microorganisms account for approximately three-fourths of all reported cases of food poisoning. Foodborne disease has reportedly become the number one food safety concern among consumers and regulatory agencies. These agencies include the U.S. Food and Drug Administration (FDA), the U.S. Department of Agriculture (USDA), and the CDC. This trend is probably true of all countries, whether developed or developing. Other terms you might hear that refer to foodborne illness are *foodborne disease, foodborne infection, food intoxication*, and *food poisoning*.

Say you're a medical doctor who suspects that your patient has contracted a foodborne disease. The first thing you should do is to inform your local or state health officials of the incident. These local health agencies have the responsibility to conduct a preliminary investigation. If they suspect the cause of illness to originate in food, they will report the incident to the appropriate federal agency that then has the responsibility to begin an epidemiological investigation. If the suspect food is still available, the investigators will normally take samples from the food, the environment, and the patient. They will then check these samples for disease-causing organisms (referred to as pathogens), for microbial and nonmicrobial toxins, and for chemicals. These test results will hopefully provide evidence of the pathogen, toxin, or chemical associated with the disease. Besides this testing, the federal agency interviews the patients—well, really, victims—and other people who consumed the same food to establish an indirect relationship of the likely foods causing the disease. U.S. federal agencies define a disease outbreak as the situation when two or more people become sick with the same illness from the same source *and* the epidemiological evidence points to the same food source. In special cases like botulism and chemical poisoning incidents, however, it takes only a single case to constitute an outbreak.

The CDC reported that between the years 1998 and 2014, there were 18,211 confirmed foodborne disease outbreaks in the United States. These outbreaks involved 358,391 individuals. Of these cases, 13,715 victims required hospital stays and 318 people died.[26] These numbers may appear low for the seventeen-year time span, but the estimated number of cases, hospitalizations, and deaths are thought to be considerably larger. For 2011 only, the CDC estimated that about 48 million people would get sick from consuming tainted foods. They linked these estimates to thirty-one different bacterial, viral, and fungal pathogens. The top five foodborne pathogens as ranked by the CDC are as follows: noroviruses, *Salmonella*, *Clostridium perfringens*, *Campylobacter*, and *Staphylococcus* (Figure 9). For hospitalizations and death, *Salmonella* led the list at 35 percent and 28 percent, respectively.[27]

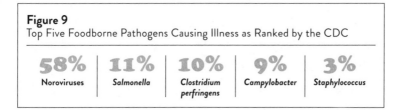

Figure 9
Top Five Foodborne Pathogens Causing Illness as Ranked by the CDC

58%	11%	10%	9%	3%
Noroviruses	Salmonella	Clostridium perfringens	Campylobacter	Staphylococcus

The discrepancies we see between the actual number of reported cases and the CDC estimates are related to several factors. If you're like us, you probably don't always go to the doctor when you are ill, so your case would go unreported. Does that sound familiar? Many of us also self-diagnose our illnesses as just the old 24-hour flu bug. It is estimated that only a small fraction of incidents are ever reported. Even if you do go to the doctor, he or she may never report the event. Finally, if your case is reported to the local health agency, it may be dropped at that point and never reach the federal level.

Foodborne illness is a costly business. Besides your individual suffering, and who can put a value on that, every incident of foodborne disease has a significant price tag. In 2012 alone, the cost of foodborne illness was estimated at $77 billion, which included medical expenses, productivity losses for being out of work, and illness-related death. This estimate did not account for the costs to the food industry, including reduced consumer confidence in the product or restaurant, product recall losses, and lawsuits, nor did it include the cost to public health agencies responding to these outbreaks.[28]

Take for example the *E. coli* outbreak involving the popular Chipotle restaurant chain in late October of 2015. We suspect many of you remember the outbreak, when about sixty customers got sick and twenty-two victims were hospitalized after eating at Chipotle. Imagine how this outbreak affected the company's bottom line and jolted consumer confidence. It crushed the "healthy food" image and scared away many customers. That one incident resulted in a 44 percent drop in the company's profits over the last quarter of 2015. Sales dropped 7 percent and Chipotle stock value declined 37 percent, an $8 billion hit compared to August of that same year. Besides sustaining these losses, the company probably paid millions of dollars more to settle lawsuits over victims' lost wages, medical bills, and pain and suffering.[29]

As for food recalls, in 2016 alone there were 764, in which the two main culprits were undeclared presence of allergens (305) and *Listeria* contamination (196).[30] Besides *Listeria*, other food recalls involving bacterial contamination included *Salmonella enterica* serotypes (99) and *E. coli* (31). Bacterial serotypes are based on cell wall surface structures that determine how these bacteria interact with their environment and how virulent they are. *Salmonella* has over 2,600 different serotypes. In 2012, due to *Salmonella* contamination that caused 41 people to become ill, the largest producer of organic peanut butter in the United States (Sunland) had to issue a twenty-state recall of all products that had been produced from March 2010 to September 2012 at its main plant in New Mexico. These products were sold in well-known stores such as Trader Joe's and Whole Foods Market. If you're thinking these recalls didn't cause these companies some headaches as well as lost revenue, think again. A company could even go out of business after a foodborne disease outbreak has involved their product.

Thanks to the peanut butter recall just mentioned, Sunland filed for bankruptcy in 2013 with an estimated $10 million to $50 million in assets and $50 million to $100 million in liabilities. A joint industry study by the Food Marketing Institute and the Grocery Manufacturers Association reported that the average cost of a recall for a food company has been estimated at $10 million in direct costs, in addition to brand damage and lost sales.[31] Direct costs typically include notification of regulatory bodies, supply chain, and consumers; and product retrieval, storage, and destruction. Add to that the labor costs associated with these activities as well as the investigation of the root cause. To appreciate the economic costs of these recalls, consider these estimates: in 1992, U.S. companies lost $160 million for *E. coli* in hamburgers; in 2006, they lost $350 million for *E. coli* in spinach; in 2007, they lost $133 million for *Salmonella* in peanut butter; in 2008, they lost $250 million for *Salmonella* in tomatoes; and in 2009, they lost $70 million for *Salmonella* in peanut products.[32]

With today's 24/7 media blitz and the capacity of social networking to rapidly spread the news, it's hard to imagine that the public is not informed and reacting to foodborne disease outbreaks and recalls. For instance, in a 2012 Harris Interactive poll, 55 percent of consumers who responded said they would switch brands temporarily following a recall, 15 percent said they would never purchase the recalled product, and 21 percent said they would avoid purchasing any brand made by the manufacturer of the recalled product.[33]

So, what kinds of foodborne diseases can affect us? There are three types of microbial foodborne illness based on how they make people sick. These illnesses result from consuming food that is contaminated with either toxins produced by toxigenic bacteria or molds or viable pathogenic bacterial cells and spores. We call the first type intoxication, because the illness results from ingesting preformed toxins that are produced in the food by some bacteria and molds. To get sick, you do not need to consume live bacteria or mold cells—their toxins alone will do the trick. (Effective little toxins, aren't they?) The two bacteria most commonly cited in this category are *Staphylococcus aureus* and *Clostridium botulinum*, the causative agents of staph poisoning and botulism, respectively. *Aspergillus flavus* is the typical mold capable of producing mycotoxins and causing a poisoning syndrome.

The second type of foodborne illness we refer to as an infection. This happens when you've consumed foods containing live pathogenic bacteria capable of infecting and growing in your gastrointestinal tract, thereby causing illness. Many bacteria and several viruses fall in this category. Some of the more common infectious bacteria are multiple serotypes of *Salmonella enterica, Campylobacter jejuni*, enteropathogenic (intestinal illness) *Escherichia coli*, and *Listeria monocytogenes*. The pathogenic enteric viruses include hepatitis A virus and noroviruses. Biologists know that unlike bacteria, which can grow and multiply on food tissues, viruses must first invade your own living cells before they can grow and reproduce. Nevertheless, some viruses can survive outside host cells for a relatively long time (weeks to months).

All foodborne viruses originate from the human intestine, and food is contaminated either by an infected food handler during food preparation or through contact with sewage, sewage sludge, or polluted water.[34] As you can see, using proper food-handling and preparation practices in the home or elsewhere does matter. It might also make you wonder if the cooks and food handlers at your favorite restaurant wash their hands after using the restroom. Of course, the same is true in your own home.

For very good reason, the Code of Federal Regulations requires restaurants to display the sign "Employees Must Wash Hands Before Returning to Work" in their restrooms.

The third type of foodborne illness is called toxicoinfection. This illness occurs when a person ingests large populations of viable cells of pathogenic bacteria that then form spores or die, releasing disease-producing toxins. Some examples in this group are *Clostridium perfringens*, *Bacillus cereus*, and enteropathogenic/enterotoxigenic *E. coli* strains (all can cause gastroenteritis, or stomach bug).

Most of these foodborne diseases have different symptoms. In all of them, the bacteria, virus, and toxin enter the body through the gastrointestinal tract, so the first symptoms are typically nausea, vomiting, abdominal cramps, diarrhea, and sometimes fever. Regrettably, everyone has had these symptoms at some time in their lives. Although your illness may be related to consuming contaminated food, it could also be a case of influenza or the "stomach bug" that we mentioned earlier. Sometimes it's difficult to identify what is causing the illness. The only sure way to know is to see your doctor and obtain lab tests.

While we're on this subject, did you ever wonder why you got sick while your brother who ate the same contaminated food didn't? Many factors can contribute to why one person suffered while the other didn't. Both of you ingested these cells and toxins, but that doesn't necessarily mean you and your brother will develop the same disease symptoms or severity of symptoms. As you read the following chapters, keep these contributing factors in mind. They include a person's relative susceptibility or resistance to

the pathogen. For example, your ability to fight off the illness may not be as good as your brother's. Sometimes age and overall health status play a role. Generally, the very young or old and those who are sick with compromised immune systems are more susceptible to microbial pathogens than normal, healthy individuals are. This susceptible group includes those with immune system diseases such as HIV/AIDS and organ transplant patients who are taking immunosuppressant drugs to prevent organ rejection. The amount of contaminated food consumed and the pathogen population or concentration of toxin in the food also affect a person's susceptibility to a microbial pathogen or toxin. In the case of you and your brother, perhaps you had a second helping of the contaminated food while he had a single helping.

A final contributing factor is the relative virulence. The more pathogenic or infective the organism is, the more severe the illness will be. For some pathogens like enteropathogenic E. coli, which infects the human gastrointestinal tract, ingesting fewer than ten viable cells can make infants or the elderly fall ill. In contrast, as a relatively healthy adult, you may need bacterial populations of a million or more viable cells for the symptoms to develop.

Any of these factors might cause individuals like you (some of whom could be involved in the questionable food-handling practices described in this book) to get sick. We think the question then comes down to knowing the probability of risk versus benefit for those who engage in these food-handling practices. Calculating actual risk probabilities for our actions is not easy, given the many factors that influence our chances of getting sick.[35]

To illustrate how age can influence people's risk of becoming ill, we want to share a personal story. Brian's wife's great-nephew Robert, who was

8 years old at the time, contracted a case of enterohemorrhagic colitis from drinking *E. coli*–contaminated water at his grandparents' farmhouse in Southwest Virginia. The water came from a nearby springhouse that happened to be located close to a field where the grandparents' milk cows grazed. Springhouses, or "natural refrigerators," were common during the late 1800s and early 1900s in North America. The springhouse was built over a spring that kept the small building and any food stored there cool. Apparently the springwater had become contaminated with *E. coli*–infected feces from the cows. His grandparents had been drinking the same water for years and never become sick, but Robert came down with this nasty illness. Fortunately, the story has a good ending. His mother, who is a nurse, quickly recognized the bloody diarrhea symptoms and immediately got him to the hospital, where he was started on antibiotics. As you can see from this real-life drama, age and building a strong immune system by being constantly challenged with contaminated water were important factors in determining why Robert got sick and his grandparents didn't.

Another significant consideration to bear in mind is that the quality of the food in question may appear perfectly fine. The food may contain viable pathogenic cells or their toxins, but that doesn't necessarily mean the quality of the food is compromised. In other words, foods may look, smell, and taste fine and have normal textures while containing pathogens and toxins at relatively low or even high concentrations. Regrettably, this means people like us may unknowingly ingest contaminated foods that otherwise appear perfectly normal. As we commented in the section on intrinsic and extrinsic factors controlling microbial growth, a single viable cell of a pathogenic bacteria can multiply in a relatively short time under favorable growth conditions. Remember that summer picnic we mentioned earlier, where the chicken casserole sat out at 90°F for 4 hours? Pathogenic viruses, on the other hand, need host cells to grow and replicate, so they cannot grow in foods and will have no impact on product quality.

In closing, we hope this introduction to the microbial world will whet your appetite to explore other aspects of the fascinating world of micro-organisms. More important, as you read through the upcoming chapters, you can decide for yourself: Is it worth taking a risk to participate in an activity that may end up being very costly? After all, isn't taking risks a part of life? At least after reading this book, you'll go into those activities with your eyes open. Just ask yourself, "Do I feel lucky?" We hope you will be, and that you'll find the remaining chapters entertaining and enlightening.

Surf

his part of the book (actually, most of the book) reminds us of our good friend Pat. Like the little kid Mikey from the Quaker Oats commercial for its Life Cereal brand, Pat is our litmus test to determine if something is okay to eat. Everyone probably has a similar friend who is always pushing the envelope when it comes to food safety. Microorganisms, as we now know, like to live on surfaces and thrive in moist environments. Microbiologists categorize bacteria as either sessile, meaning attached to a surface like a barnacle, or planktonic, which refers to organisms that flow around freely like the fish swimming in a lake.

As it turns out, bacteria mostly grow on surfaces as "communities," also known as biofilms. Although our old friend Antonie van Leeuwenhoek saw biofilms in the 1600s, real progress in understanding them began with a 1978 report entitled "How Bacteria Stick."[1] To form a biofilm, bacteria first attach to a surface. They multiply and then produce structures called polysaccharide layers, which are kind of like buildings made from complex sugars. These bacterial buildings can grow from ten to fifty times the length of one bacterium.[2] The bacteria in a biofilm now have shelter to protect themselves from changes in the environment, like drying out, as well as from attacks by humans such as the use of sanitizers and antibiotics. These biofilms are so organized that bacteria take on different roles to help sustain their "bacteria town."

How do they do this? Believe it or not, they do communicate. Bacteria release chemicals known as acylated homoserine lactones. They use these chemicals like people use words. When the chemicals reach other bacteria, genes are triggered to give a desired response.[3] This bacterial communication is called quorum sensing because it needs to take place in a bacterial population large enough to produce the amount of chemicals necessary to communicate that it's time for the bacteria to act as a team. When bacteria have physical protection and are communicating in this way, they are extremely difficult to eliminate and can be dangerous to humans. One problem all people have in common is tooth biofilms, called plaque, that can lead to more serious issues such as gingivitis or gum disease. Yeah, that coating on your teeth each morning is an overnight biofilm that you can't remove without brushing.

Bacteria on surfaces that form biofilms can also cause serious diseases, including conjunctivitis, colitis, vaginitis, and otitis. Biofilms are a significant problem primarily because (1) they allow bacteria to be highly resistant to antibiotics, sanitizers, and nearly anything that kills free-flowing cells; and (2) they periodically release cells that can then infect other locations within the body or the environment. Bacterial biofilms create serious problems for medical devices and implants. Intravenous tubes, prosthetic implants, heart valves, pacemakers, respirators, contact lenses, and blood vessel stents can form biofilms that result in recurring and systemic infections.

Biofilms also play an important role in the genetic disease cystic fibrosis. Among those of European descent, it is the most prevalent lethal genetic disease, affecting 30,000 children and adults in the United States. Cystic fibrosis is characterized by a progressive loss of lung function resulting from chronic bacterial infection. And the strong biofilm-forming capability of the infecting bacteria is a major reason that the bacterial infection resists treatments.[4]

So what about the surfaces that we come into contact with every day? These surfaces, as you might have guessed, are not clean. One study looked at 1,061 public places including stores, day care centers, offices, gymnasiums, airports, movie theaters, and restaurants located in the cities of Chicago, Tucson, San Francisco, and Tampa. The researchers were looking for biochemical markers such as blood (hemoglobin), mucus (amylase), and urine/sweat (urea) as well as protein, total coliforms (bacteria commonly found in soil, runoff, and excrement), and fecal bacteria, all of which are general hygiene indicators. Twenty-one percent of movie theaters and 51 percent of restaurants had surfaces judged to be highly contaminated (meaning they were visibly soiled and had more than 200 micrograms of protein in an area measuring 10 square centimeters).[5] One in four of all the surfaces tested were judged to be heavily soiled (> 200 μg protein per 10 cm^2), while one in five tested positive for at least one biochemical marker (blood, urine, sweat, or mucus). Yuck!

Cell phones are reservoirs of bacteria, too. One study found that 75 percent of the phones sampled in health-care settings had at least one potential disease-causing pathogen.[6] (Think about that the next time you make a call.) And for surfaces in your home, WebMD listed the dirtiest places in the home as the TV remote, salt and pepper shakers, toothbrushes, computer keyboards, and bathtubs/whirlpools.[7] Whirlpools were especially dirty—81 percent contained fungi, 34 percent contained staph, and nearly all had fecal matter present! Whoa. Flash back to the "doodie in the pool" scene from the movie *Caddy Shack*. We're not talking about a Baby Ruth bar here. But unlike Carl Spackler (Bill Murray), we do think it's a big deal. Common household surfaces are our focus in this part of the book as we address the five-second rule and dropping food on tile, wood, and carpet; beer pong where the surface is both the Ping-Pong ball and all of the other surfaces the ball bounces on during the game; and restaurant menus. So let's dive into our first myth, the five-second rule, and see what Genghis Kahn and Julia Child have to do with it.

THE
FIVE-SECOND
RULE

Chapter 1

THE FIVE-SECOND RULE

Suppose you just dropped a piece of Godiva dark chocolate on the floor and quickly picked it up. Your mind instantly
starts weighing your options, the pros and cons of eating the
chocolate—the angel on one shoulder and the devil on the other. The
good angel tries to convince you to discard the chocolate because it may
have picked up potentially dangerous bacteria and you may get sick—or
worse. The devil responds that it doesn't even matter if there are dangerous bacteria on the floor, because you can apply the infamous five-
second rule.

Known as the five, ten, (you fill in the blank)–second rule, this urban
myth proposes that if food is removed from a contaminated surface
quickly enough, the microorganisms on the surface won't have time
to transfer, or "jump," onto the food. In the food production and ser-

vice industries, it's common practice to throw away food intended for human consumption if it has been dropped onto unsanitary surfaces. However, there is a perception that if dropped food is picked up quickly enough from a dirty surface, the food may be "okay" to eat.[1] In fact, some scientists have postulated that allowing children to practice the five-second rule might improve their immune systems.[2] Yeah, and drinking from the sewer might also improve your immune system—if you survive. Some urban myths have a scientific basis, whereas the origins of others are simply unknown. Are seconds a critical time frame that determines if food is safe to eat, or aren't they? As you'll see, this food myth is affected more by how dirty the floor is than by how long the food lies there.

ORIGIN OF THE FIVE-SECOND RULE

The rules about eating food off the floor are sometimes attributed to Genghis Khan (1162–1227), who is said to have instituted the "Khan Rule" at banquets for his generals: If food fell on the floor, it could stay there as long as Khan allowed, because food prepared for Khan was so special that it would be good for anyone to eat no matter what.[3]

In reality, people had little basic knowledge of microorganisms and their relationship to human illness until much later in our history. Thus, eat-

One Genghis Khan, Two Genghis Khan, Three Genghis Khan . . .

ing dropped food was probably not taboo before we came to this understanding. People could not see the bacteria, so they thought wiping off any visible dirt made everything fine.

Television's original queen of the culinary arts, Julia Child (1912–2004), may have contributed to the urban myth by routinely picking up dropped food while preparing delectable meals. Jessie Schanzle, a writer for the news website "The Conversation," discovered that a well-known but inaccurate story about Child had contributed to the myth.[4]

Viewers of her cooking show, *The French Chef*, have claimed they saw Child drop lamb (some say chicken or a turkey) on the floor and pick it up, advising viewers that if they were alone in the kitchen, their guests would never know. In fact, it was a potato pancake that fell onto the stovetop, not on the floor. Child returned it to the pan saying, "But you can always pick it up, and if you are alone in the kitchen, who is going to see?"[5] But in popular culture the misremembered story persists.[6]

FIVE-SECOND-RULE STUDIES

Tests of the five-second rule have been presented on several television shows, in academic news releases, and in only two published research studies—one of which was conducted in our laboratory.[7]

The first research study directly addressing the five-second rule was announced in a 2003 press release from the University of Illionis.[8]

In that study, gummy bears and fudge-striped cookies were dropped on vinyl floor tiles inoculated with *E. coli*. The *E. coli* was transferred from the tiles to the gummy bears and fudge-striped cookies within five seconds, but the researchers did not report on the number of bacteria transferred.

Based on surveys of college students, the authors also found that the gender of the person eating off the floor and the type of food on the floor affected the chance of eating food off the floor:

- Seventy percent of females and 56 percent of males were familiar with the five-second rule, and most of them use it to make decisions about tasty treats that slip through their fingers.

- Women are more likely than men to eat food that's been on the floor. Who would have thought? Maybe females are less wasteful than males?

- Cookies and candy are much more likely to be picked up and eaten than cauliflower or broccoli. No surprise here!

The popular Discovery Channel television series *MythBusters* got in on the act in 2005 with an episode that aired on October 19.[9] After conducting a few tests, the show's hosts, Jamie Hyneman and Adam Savage, concluded that contact time (2 or 6 seconds) was not a determining factor in the transfer of bacteria to food. They dropped wet food (pastrami) and dry food (crackers) onto contaminated surfaces and found that the pastrami picked up more bacteria than the crackers did. Jamie admitted that they needed to run more tests to see if there was a difference between 2 and 6 seconds. As a mini-test, they placed contact petri dishes at various locations around the shop and found that toilet seats were cleaner than floors. Like most *MythBusters* episodes, these were not statistically designed research studies. What does *cleaner* even mean without some kind of context or control group?

In 2006, our Clemson University study was the only scientifically peer-reviewed paper on this topic published to that date.[10] We investigated whether the length of time that food is in contact with a contaminated surface does in fact affect the transfer of bacteria to the food. You'll find more details on this study later in this chapter. Briefly, though, here is the process: We (1) inoculated square samples of tile, carpet, and wood with a strain of salmonella; (2) dropped food on these surfaces; and (3) then measured the number of bacteria transferred from the surface to the food.

One year later, in 2007, two undergraduate microbiology students at Connecticut College reported that Skittles were safe to eat after 30 seconds of contact with the floors located in the university dining hall and snack bar, whereas apple slices were safe to eat after more than 1 minute.[11] Since the levels of surface contamination (including pathogens) were not reported, their results most likely depend on whether the surface was contaminated or not. The article "To Eat or Not to Eat: Seniors Prove 'Five-second Rule' More Like 30" concluded that "no bacteria were present on the foods that had

remained on the floor for 5, 10 or 30 seconds." The conclusions of this report are somewhat confusing and contrary to nearly all other reported studies focused on bacterial adhesion to wet and dry food surfaces. In a 2014 press release, researchers from Aston University in Birmingham, England, reported that contact time significantly affected the transfer of *E. coli* and *Staphylococcus* from inoculated carpet, tile, and laminate to toast, pasta, and sticky candy.[12] They also reported that 87 percent of people they queried either would or have eaten food dropped on the floor.

In January 2016, the Science Channel's television series *The Quick and the Curious* showed NASA engineer Mick Meacham offering cookies to strangers after dropping them on the ground in a park. The show's host stated that moist foods left on the ground for 30 seconds picked up ten times more bacteria than moist foods left for only 3 seconds, but no data or tests with moist foods were reported to support this statement.[13]

In 2016, nine years after the Connecticut College study, a second peer-reviewed paper was published on the topic from Rutgers University. The results were similar to those reported in our 2006 study, although the researchers included a wider range of foods and bacteria.[14] They tested watermelon, bread, bread with butter, and gummy bears on tile, stainless steel, wood, and carpet for 0, 5, 30, and 300 seconds. They found that bacteria transfer to food ranked from highest to lowest in the following order:

watermelon > bread = bread with butter > gummy bears

Generally, though, they observed the same trends reported in our 2006 study.

SO, WHAT DOES THE RESEARCH TELL US ABOUT THE FIVE-SECOND RULE?

Can you drop food on the floor, pick it up, and then eat it with no risk of ingesting pathogenic microorganisms? Until recently, only one peer-reviewed research study actually tested the rule. Then an even more recent publication pretty much nailed down the answer. The five-second rule appears to be an old wives' tale. The differences in the conclusions drawn from these previous studies are attributed to how the studies were designed and conducted. For example, the Connecticut College researchers applied their test to a real-world scenario by choosing surfaces at the university where people dine. Were they really testing the five-second rule, or were they testing the chance that the surfaces would be contaminated? We believe they were testing the latter, since they didn't appear to determine the contamination level of the surface on which the food was being dropped.

There is conclusive evidence that when food comes into contact with a contaminated surface, bacteria are transferred almost immediately. Eating food that has been dropped on the floor could be compared to driving a car without wearing a seat belt. You could drive your whole life without wearing a seat belt and never have an accident, but it doesn't prove that wearing a seat belt won't prevent injury in case of an accident. Similarly, eating food from a non-

"Click It" or Get Sick

contaminated surface poses no risk. However, many factors affect the associated safety risk of eating food that has contacted a surface. They include but are not limited to the dose/population and type of microorganisms present, presence of pathogenic or nonpathogenic organisms, composition/characteristics of the contact and microorganism surfaces (charge, hydrophobicity, etc.), and general health status of the consumer.

OUR STUDY

Let's say you're making a bologna sandwich. If you put the bologna on the tile counter or somehow drop it on the floor, is it okay to pick it up and use it in the sandwich? What about using bread that has been lying directly on the counter? To investigate this scenario, we conducted two experiments: In the first one, we wanted to find out if food picked up bacteria when dropped on tile, carpet, or wood flooring; in the second, we wanted to see how long bacteria could survive on one of the test surfaces (tile). In the second experiment, we also wanted to know how long a surface would remain "hot" and thus give innocent five-second-rulers a chance to unknowingly test their immune systems.

EXPERIMENT 1-1: Is Salmonella Transferred in 5, 30, or 60 Seconds?

Materials and Methods

We dropped slices of bologna or bread on tile, carpet, and wood flooring that had been inoculated with *Salmonella* Typhimurium, an organism

responsible for many foodborne disease outbreaks. With genome sequencing the technical
name for this bacterium is *Salmonella enterica*
subspecies *enterica*, serotype Typhimurium, but
from now on we will shorten the name to *S. Typhimurium*. The food was left on the surface for 5,
30, or 60 seconds before we picked it up and ran
a test to estimate the number of bacteria it had
acquired. Additionally, we determined the population of salmonella on bologna and bread after
allowing it to dry for 5 minutes, 2 hours, 4 hours, 8
hours, or 24 hours.

Science Stuff Ahead

Samples of the three contact surfaces under investigation were first
sterilized in an autoclave (a large pressure cooker designed to kill all microorganisms). Then each sample was inoculated with a pure culture of *S. Typhimurium*. The *S. Typhimurium* environmental isolate used in this study was
resistant to 1,000 parts per million (ppm) of nalidixic acid (an antibiotic),
which facilitated the recovery and enumeration of salmonella populations
without interference from other naturally present bacteria. A new inoculation culture was prepared for each of the three experimental replicates.

To prepare the surfaces being inoculated, cells of a 24-hour culture
of *S. Typhimurium* were harvested by centrifugation and then resuspended in 0.1 percent sterile peptone water to obtain an approximate
working concentration of 10^{7-8} CFUs/mL. The abbreviation CFU refers to
colony-forming unit, or the number of colonies that were counted on a

petri dish per some plated volume such as milliliters (mL). One colony is presumed to have originated from a single bacterium. One milliliter of a 10^{7-8} CFUs/mL cell suspension was spread in a circular motion onto each of three 10×10 cm surfaces (tile, wood, carpet) using a sterile L-shaped glass rod.

After air-drying each surface for 5 minutes, the investigators inoculated samples of tile (glazed ceramic tile by American Olean), Berber-style carpet (ST103 Stratos, sold by Lowe's), and wood (Bruce pre-finished polyurethane hardwood floors, Dura-Luster Plus urethane finish by Armstrong Co.). Each sample was held in an environmental chamber at 70°F and 50 percent relative humidity for residence times of 5 minutes and 2, 4, 8, and 24 hours before sampling. At each of the sampling times, bologna slices measuring 10×10 square centimeters (cm^2) were aseptically placed onto the inoculated surfaces, then aseptically removed after 5, 30, or 60 seconds of contact time. Cells were recovered from each bologna slice by rinsing the surface with 10 mL of 0.1 percent peptone water in a sterile Stomacher bag followed by 30 seconds of massaging within the bag. These same procedures were followed for testing white bread on inoculated ceramic tile only.

Statistical Analysis

Each surface was treated independently in the statistical analyses, and each experiment was replicated three times on different days (analyzed in duplicate per replication) using new contact materials. Main effects—residence time and bologna (or bread)—and their interactions were tested to determine statistical difference at a significance level of 5 percent using the Statistical Analysis System (SAS).[15]

DID YOU JUST EAT THAT?

Results of Experiment 1-1

Well, our findings pretty conclusively busted the myth of the five-second rule. We found that the bacteria are transferred to the bologna after only 5 seconds of contact time, thereby demonstrating that they may not be safe to eat. However, the statistical analysis revealed that more bacteria were transferred after 30 and 60 seconds than after 5 seconds (Figures 1 and 2). This is why some say the five-second rule is true.

(See Figures 1 and 2.)

The longer the bacteria were allowed to stay on surfaces before applying the bologna, the lower the population of S. Typhimurium transferred to the bologna (Figures 1 and 2). Several factors may contribute to why these populations on the bologna diminished as incubation time increased. First of all, cells can die off due to the lack of essential nutrients for survival on the tile, wood, and carpet. (Remember how bacteria need nutrients, too?) A second factor may be greater physical attachment of the cells to the three contact surfaces, the longer the bacteria were allowed to remain on the surfaces. Greater physical attachment could mean that fewer cells would be transferred to the bologna. And finally, the different surface area of materials could have contributed to the variations in population recovered from the three surfaces. Presumably the carpet would have the greatest surface area, followed by the smoother wood and tile surfaces.

You'll notice that these graphs show the salmonella populations listed as log base 10 ($\log_{10}$) values. Bacterial populations are often reported in this way because they are found in such high numbers, and they grow—and die—in a logarithmic or exponential pattern. An easy way to mentally convert $\log_{10}$ values into conventional numbers is to think that the log value is equivalent to the number of digits following the first number: for instance, 6 logs = 1,000,000 (adding six zeros to the right of 1).

We found some noticeable differences between carpet, wood, and tile. After 5 minutes of incubation, fewer bacteria were transferred to

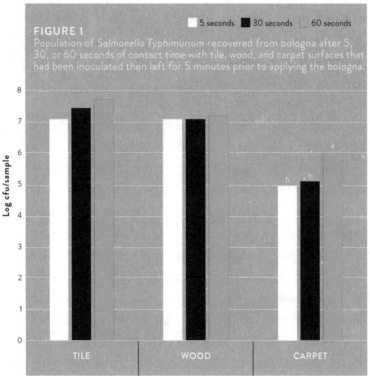

FIGURE 1
Population of *Salmonella Typhimurium* recovered from bologna after 5, 30, or 60 seconds of contact time with tile, wood, and carpet surfaces that had been inoculated then left for 5 minutes prior to applying the bologna.

a,b Bars with different letters within each incubation/hold time are significantly different at the 5% level. Values for residence times having no letters were not statistically different at the 5% level.

the bologna from the carpet than were transferred from the tile or wood (see Figure 1). After 24 hours of incubation of the bacteria on the three surfaces, the bologna samples acquired significantly more bacteria from the tile than from wood or carpet (see Figure 2). What's more, based on how long the bologna stayed on the surfaces (5, 30, or 60 seconds), we observed some statistically significant differences in the number of bacteria transferred. In general, we found that longer contact time yielded higher numbers of bacteria on bologna.

Hey buddy!

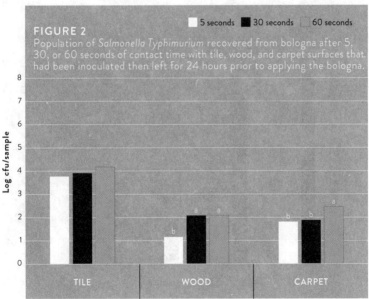

FIGURE 2
Population of *Salmonella Typhimurium* recovered from bologna after 5, 30, or 60 seconds of contact time with tile, wood, and carpet surfaces that had been inoculated then left for 24 hours prior to applying the bologna.

■ 5 seconds ■ 30 seconds □ 60 seconds

a,b Bars with different letters within each incubation/hold time are significantly different at the 5% level. Values for residence times having no letters were not statistically different at the 5% level.

So what about that sandwich bread? You might think that because bread is drier than bologna, it would pick up fewer bacteria when you drop it on a surface like ceramic tile. Not so fast. We found that bread acquired significant levels of *Salmonella* cells at all five of the hold times (0–24 hours). The longer the incubation time before applying the bologna, the lower the population of *S. Typhimurium* transferred to the bologna (Figure 3). Several factors may contribute to why these populations on the bolo-

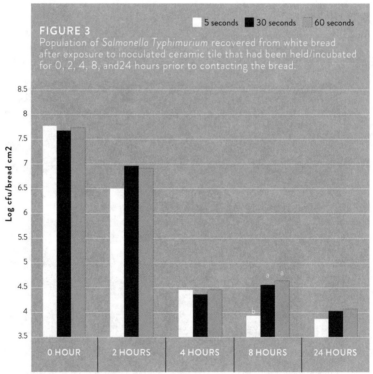

FIGURE 3
Population of *Salmonella Typhimurium* recovered from white bread after exposure to inoculated ceramic tile that had been held/incubated for 0, 2, 4, 8, and 24 hours prior to contacting the bread.

a,b Bars with different letters within each incubation/hold time are significantly different at the 5% level. Values for residence times having no letters were not statistically different at the 5% level.

gna diminished as incubation time increased. The first factor is that the cells can die off due to the lack of essential nutrients for survival on the tile, wood, and carpet. (Remember how bacteria need nutrients, too?) The second factor may be an increase in the physical attachment of the cells to the three contact surfaces as incubation time increased. Greater physical attraction could mean that fewer cells would be transferred to the bologna. And finally, the surface area of these contact surfaces could have contributed to the variations in population recovered from the three surfaces. Presumably the carpet would have the greatest surface area, followed by the smoother wood and tile surfaces. A similar trend of reduced transfer at longer hold times was found for bologna at these five hold times.

EXPERIMENT 1-2: Is Salmonella a Bad Houseguest?

Materials and Methods

Because food might drop onto a surface that has been sitting for a while after someone else contaminated it, we decided to see how long *Salmonella* would survive on ceramic tile for up to 1 month. Ceramic tile was first sterilized, inoculated with *Salmonella Typhimurium* suspended in one of three different solutions containing varying concentrations and types of nutrients, and then air-dried for 5 minutes. The inoculated tiles were placed in a temperature- (70°F) and humidity-

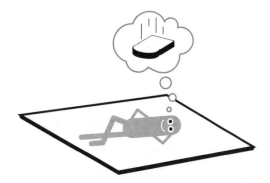

(50 percent) controlled chamber, and populations of surviving bacteria were determined after 1, 2, 3, and 4 weeks. *Salmonella* cells were suspended in these three different nutrient-rich solutions to determine if this helped them survive. We chose this experimental design because it more realistically mimics a real-life scenario where contact surfaces are often contaminated with food debris and liquids.

Science Stuff Ahead

The *S. Typhimurium* environmental isolate was used as described in Experiment 1-1, except the sterile (autoclaved) tile (finished ceramic tile by American Olean) surfaces (10 × 10 cm) were inoculated with 1 mL of the log 7–8 CFUs/mL suspension of *S. Typhimurium* that had been suspended in either 0.1 percent peptone water, 1.0 percent tryptic soy broth (a nutritious medium used for growing aerobic bacteria), or 10 percent tryptic soy broth. Cells were again recovered from each tile by rinsing the surface with 10 mL of 0.1 percent peptone water in a sterile Stomacher bag, followed by 30 seconds of massaging within the bag. Recovered bacteria were enumerated by serial dilutions in 0.1 percent peptone water and then pour-plated in duplicate using tryptone soy agar (Difco Laboratories) containing 1,000 ppm nalidixic acid. Plates were incubated for 48 hours at 70°F, after which plates containing from 25 to 250 cells per plate were counted and the counts converted to CFUs/cm^2 and log CFUs/cm^2. Each sampling time was conducted in duplicate over three replications, resulting in six observations per sampling time.

Results of Experiment 1-2

Like a bad houseguest, *Salmonella* hung around for a month (28 days) after being invited to the tile surface (Figure 4). The tile seemed completely dry and showed no visual indication that the surface carried live bacteria within about an hour. However, between 3,000 and 30,000 *Salmonella* were found on the tile after 28 days. The higher population was found on tile where the *Salmonella* had more food to snack on (10 percent tryptic soy broth). The tile was inoculated with 100 million bacteria, which is a lot, but is what a highly contaminated surface might naturally carry. Since viable *S. Typhimurium* cells were recovered at significant levels after 28 days of incubation, it is likely that they could survive even longer.

You might be shocked by this finding, but it is not unexpected in the microbial world. Bacteria capable of forming spores (that is, *Clostridium* and *Bacillus* species) are known to survive for years in their

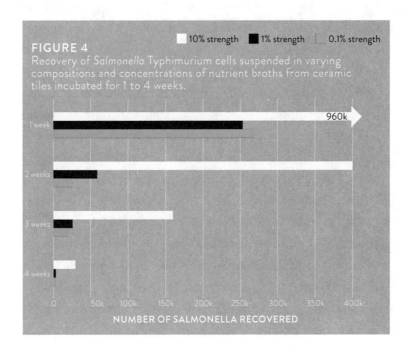

FIGURE 4
Recovery of *Salmonella* Typhimurium cells suspended in varying compositions and concentrations of nutrient broths from ceramic tiles incubated for 1 to 4 weeks.

10% strength · 1% strength · 0.1% strength

960k

NUMBER OF SALMONELLA RECOVERED

dormant spore form. In fact, spores (specifically endo-spores) found in the tombs of Egyptian pharaohs have germinated after being dormant for thousands of years, and bacterial spores from bees trapped in amber for 25–40 million years have been regenerated.[16] The terms *spore* and *endospore* are often used interchangeably even though they have different meanings. Endospore describes a spore produced inside a "mother bacteria," while spores are technically active, reproductive structures sometimes associated with the life cycles of various plants, algae, fungi, protozoa, and bacteria. Even bacteria incapable of gener-ating these resistant spores can survive on dry surfaces for days to weeks. Other studies found that *Salmonella* species could persist on a laminated countertop surface for at least 24 hours. *E. coli* and *Salmo-nella* species were also found to survive on clothes, hands, and utensils for up to two days.[17]

As you may know, *Salmonella* species are common foodborne patho-gens associated with raw foods and implicated in cross-contamination scenarios. The nutrient concentration on a surface carrying bacteria will affect how many bacteria survive on that surface. In our study, the full-strength growth medium diluted to 10 percent retained more bacteria than the 1 and 0.1 percent media did over 4 weeks (see Figure 4).

Fun fact: *Salmonella* likes carpet. We also tested how *S. Typhimurium* tolerated tile, wood, and carpet over 24 hours using the 0.1 percent pep-tone water solution to inoculate the surfaces. Over the first 24 hours, the population of salmonella remaining on the three inoculated surfaces was significantly higher on carpet as compared to tile and wood (Figure 5). We can probably attribute this finding to the larger available surface area for bacteria to reside on and a slower drying time.

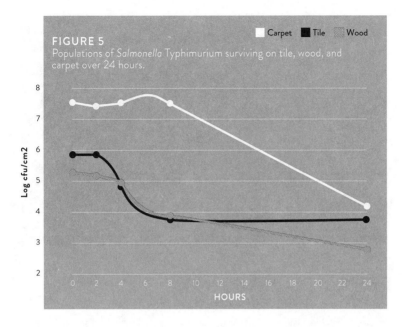

FIGURE 5
Populations of *Salmonella* Typhimurium surviving on tile, wood, and carpet over 24 hours.

■ Carpet ■ Tile ▨ Wood

Another interesting finding was that carpet had a very small amount of bacterial transfer (< 0.5 percent) compared to tile and wood (5 to 69 percent; Table 1). However, thanks to the greater survival of *Salmonella* in contact with carpet, some bacterial transfer did occur throughout the 24-hour sampling time. Carpet offers a larger surface area for bacteria to grow on because they can attach to the woven fibers; tile and wood have smaller surface areas. In the earlier contact times, the bacterial cells may have been more distributed throughout the carpet so the food contacted only a small fraction of the total carpet surface area and resulted in less transfer.

Bacteria can reside on surfaces for some time before being transferred. For example, *S.* Enteriditis survived on laminated countertop for 24 hours after being exposed to contaminated egg. Other studies have

TABLE 1
Percentage of *Salmonella* Typhimurium transferred to bologna and bread from different contact surfaces. The percent transfer was calculated from the population of bacteria remaining on the surfaces at each residence time to the initial inoculum population.

RESIDENCE TIME	BOLOGNA ON WOOD	BOLOGNA ON TILE	BOLOGNA ON CARPET	BREAD ON TILE
HOURS		% TRANSFER		
0	47.9	68.5	0.1	48.7
2	30.5	40.0	0.4	44.8
4	5.5	14.7	0.1	5.7
8	13.1	48.6	0.02	16.1
24	9.7	25.9	0.2	32.7

reported that bacteria are transferred from stainless steel to cucumber,[18] from stainless steel to lettuce,[19] and from cutting boards to lettuce.[20] While drying a surface will cause a reduction in bacterial populations, *Campylobacter* and *Salmonella* species (spp.) decreased only from 5.5 to 4.0 log CFUs on stainless steel after 2 hours of drying.[21]

THINGS TO CONSIDER

So, eating food off of any floor is like eating off a surface where people, pets, and possibly other animals have walked throughout the day. Think about that: It's a surface where animals have possibly relieved themselves. The average adult dog produces over 200 pounds of feces a year, and with 78–83 million dogs in the United States, that adds up to 20 billion pounds of dog doo-doo on the ground per year. Okay, so some people pick up after their dogs. What about urine? A 10-pound dog generates about

8 ounces of urine each day. With around 80 million dogs, this adds up to 5.1 million gallons a day or 1.9 billion gallons each year. Ever been to a city where dogs relieve themselves on the sidewalk? Walking on the sidewalk is like walking by a bank of urinals.[22]

Similar to how wearing a seat belt significantly reduces your risk of injury, not eating dropped food significantly reduces your risk of illness if there are pathogenic, or disease-causing, bacteria residing on the contact surface. Parents from past generations have told their offspring, "You can eat a pound of dirt and not get sick." True, exposure to a variety of microorganisms can strengthen your immune system. However, there's no denying that certain pathogenic bacteria, viruses, and parasites in our environment can make you sick.

Another frequently heard remark is that people residing in other less developed countries or cultures are typically less concerned with food sanitation, yet they still survive. This is true, but it's also true that earlier generations did not live as long as those today and that people living in developing nations often suffer frequent bouts of gastroenteritis that may be associated with consuming tainted food and water. What's more, bacteria and viruses are constantly adapting to the environment, leading to the emergence of more deadly strains. Thus, by examining the five-second rule in the lab, we have learned that the factors determining if you get sick from eating tainted food include the dose of the pathogen, the virulence of the pathogen, and how healthy you are.

BEER PONG

Chapter 2

BEER PONG:
DON'T HATE THE GAME

As professors who work with undergraduate students, we are familiar with a game called beer pong. But for those readers who may not know, beer pong is a drinking game in which players attempt to throw or hit table tennis balls into cups of beer, and their opponents are required to drink the contents of any cup in which a ball lands. This is how the Oxford English Dictionary uses *beer pong* in a sentence: "The bar was fairly empty but we had a great time playing beer pong."

The popular belief is that fraternities at Dartmouth College invented beer pong in the 1950s and then other colleges like Bucknell and Lehigh later modified it. The game was first played with paddles, and today's version using hands was originally known as Beirut.[1] Beer pong gained international recognition when the first World Beer Pong Championship was held in 2009, offering a grand prize of $50,000 to the winning team. Several local, state, and national beer pong leagues now exist: the National Beer Pong League, the World Series of Beer Pong (which holds the annual world championship each year), the European Series of Beer Pong, the Australian League, and several official state leagues in the United States.

Beer pong has many variations. However, as mentioned earlier, the basic game involves throwing a Ping-Pong ball into your opponent's half-filled cup of beverage (usually beer) from across a long, narrow table. A successful throw requires your fellow player to drink the contents of the cup. A good player has the advantage of getting opponents drunker and drunker, so their throws become less accurate. The player who makes the most successful throws wins the game, forcing the loser to drink

the beer left in the cups. Losing is no fun. This much, the participants know. But in addition to becoming inebriated, what other potentially bad things might be going on?

During a typical game, the ball bounces on the playing surface, misses a cup, and lands on the floor or the ground; from there it is recovered and returned to play. Now some people, we've heard, will rinse the ball in water between throws to "clean off" the ball. But wetting the ball like this might actually make it pick up *more* bacteria from the hands and wherever else it lands. Tracking down a runaway pong ball calls to mind the famous remark by major league baseball player Bob Uecker: "The way to catch a knuckleball is to wait until it stops rolling and then pick it up." Playing beer pong like this gives microorganisms lots of opportunity to transfer from various surfaces and onto the ball. Beer pong players should also know that pong balls are *not* sterile when removed from the package—they just appear to be relatively clean, especially as compared to how they look after being handled and bounced around.

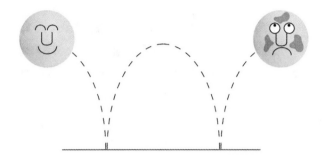

While most bacteria in the home, outside environment, or off-campus housing and basements are not necessarily harmful, we now know that some bacteria and viruses cause disease. This means handling a ball exposed to non-sterile surfaces (including hands) likely increases the risk of exposure to such pathogens. Since beer pong involves drinking beer that has been exposed to a Ping-Pong ball, our research team wanted to determine if—and how much—bacteria are transferred to beer during the playing of beer pong. Our hypothesis was that bacteria do contaminate Ping-Pong balls during beer pong, and that the bacteria on the balls are also transferred to the beer during a beer pong match. Given what we now know about pathogens, the beer pong player who drinks the contaminated beer could lose a lot more than the match.

We divided our specific objectives into two experiments: in Experiment 2-1, we wanted to determine the overall number of bacteria found on Ping-Pong balls used to play beer pong during a football homecoming weekend. Our objective in Experiment 2-2 was to run a controlled laboratory test to measure the number of bacteria transferred to beer from balls inoculated with a known number of bacteria.

EXPERIMENT 2-1: Bacteria Found on Ping-Pong Balls Used in Homecoming Beer Pong Games

Materials and Methods

To determine the numbers and types of bacteria found on Ping-Pong balls used in actual beer pong games, brave student researchers fanned out across the campus on homecoming weekend. Their goal was to infiltrate dens of beer pong activity to collect Ping-Pong balls being used in beer pong games (in exchange for unused balls). What the participants in the beer pong games thought of this research was not recorded and best not discussed.

In all, 63 beer pong balls were aseptically placed into sterile plastic bags and transported to the lab. They were collected from games played on wooden porches (12), hardwood floors (4), vinyl flooring (17), carpet (6), and outdoors on dirt and grass (24). All pong balls collected were observed to have been dropped on the ground or floor more than once per game and had rolled various distances on the ground, porch, carpet, vinyl or hardwood floors. Also, each ball was handled once per turn by each of the players and would often bounce or roll on the table or ground and would sometimes even land in the cup of beer. Drink! Drink! Drink!

Science Stuff Ahead

To recover bacteria from the Ping-Pong balls, each ball, held in a separate sterile Whirl-Pak sample bag (made by Nasco), was delivered to the lab. There, 20 mL of 0.1 percent sterile peptone water was added to each bag. The bag was then shaken for 30 seconds to remove bacteria. A milliliter of the solution used to rinse the ball was serially diluted, and then 1 mL of the appropriate dilutions was transferred in duplicate to the surface of separate petri dishes containing plate count agar. Samples were spread evenly over the plates and incubated at 37°C for 72 hours. Colony-forming units (CFUs) were counted on each set of duplicate plates containing between 25 and 250 colonies of growth. In theory, one cell divides during incubation into multiple cells, eventually forming a visible colony composed of thousands to millions of cells. Each colony therefore is counted as originating from one cell. These counts were converted to CFUs/mL of rinse and CFUs/Ping-Pong ball. Isolates from each distinct type of CFU were examined for morphology (shape and appearance) and gram stain reaction (a test that divides bacteria into two general categories based on composition of their cell walls).

Data collected from the random sampling on campus was categorized based on the location and surface on which the beer pong game was played. Thus, balls were categorized as carpet, hardwood, vinyl, outdoors, or porch. These data were subjected to a statistical analysis to generate bacterial count means (averages), standard deviations (measure of degree of variation about the mean), medians (center value), and ranges (high and low value).[2]

This experiment would determine whether the balls in play acquired bacteria from being in play.

Results of Experiment 2-1

What a bummer. Our tests revealed that numerous bacteria were indeed found on the surface of the beer pong balls. The bacterial populations varied widely according to where the game was being played (Figure 1). Remember, in Experiment 2-1 we were trying to determine the level of ball contamination found only in real beer pong games being played on a typical college campus and not in a laboratory setting. The average number of bacteria recovered from all pong balls regardless of the playing environment was 76,000. The good news? Balls from games played on carpet averaged only 600 bacteria per ball (average of six balls). The bad news is that most of the games (using an average of 24 balls) were played outdoors on grass and dirt, and the bacterial count reached a whopping average of 201,165 bacteria per ball.

First, keep in mind that this was an average count; the actual highest number of bacteria found on one ball was around three million cells! Of course it's the bacteria found on an individual ball that causes illness, not the average bacterial population. It really comes down to whether any of these bacteria are pathogenic. The types of bacteria recovered from

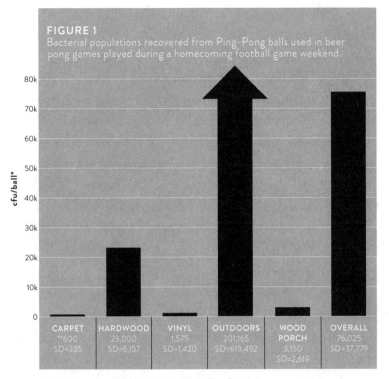

FIGURE 1
Bacterial populations recovered from Ping-Pong balls used in beer pong games played during a homecoming football game weekend.

CARPET	HARDWOOD	VINYL	OUTDOORS	WOOD PORCH	OVERALL
**600	23,000	1,575	201,165	3,150	76,025
SD=335	SD=6,157	SD=1,420	SD=613,492	SD=2,619	SD=37,779

*CFU/ball = colony forming units (bacteria) per Ping-Pong ball.
**Mean with standard deviation (SD) for each surface.
N = number of Ping-Pong balls tested per surface: carpet (6), hardwood (4), vinyl (17), outdoors (24), wood porch (12), overall (63).

Ping-Pong balls collected from the beer pong games played during the homecoming weekend were the kind typically found in soil, on hands, and generally on surfaces found in the environment. Hands or any other items that are in contact with contaminated surfaces play an important role in transmitting bacteria from surface to surface. Some bacteria found in the environment can even pose a threat to humans. Contaminated hands are a major source of cross-contamination in food-service operations, so balls handled during beer pong could potentially be a source of human pathogens.[3] Transfer of pathogens from hand to mouth is also a common cause of foodborne illness and can be reduced by minimizing the handling of ready-to-eat food with contaminated hands.[4]

So if you play beer pong, you are probably drinking beer containing bacteria found on whatever surface the ball has previously touched. Furthermore, the infectious dosage of pathogens for each of us will vary based on our personal health status, as we mentioned earlier in the introduction, "A Dive into the Mysterious Microbial World." It is clear that the infectious dose is more likely to be reached when populations of bacteria are higher, but it doesn't mean lower populations won't cause illness. Still, are you willing to take that chance?

There was a wide range in the number of bacteria found on individual balls we collected. For example, a game played outdoors yielded one ball containing nearly three million bacteria, and the fewest number of bacteria recovered from one ball was 180 in a game played in a room with vinyl flooring. Does that mean that grass is worse than vinyl for playing beer pong? Let's find out, young grasshopper.

EXPERIMENT 2-2: Do Bacteria Travel from the Ping-Pong Balls to the Beer?

Materials and Methods

Experiment 2-1 determined the number of environmental bacteria that ended up on the ball during beer pong games. Experiment 2-2 was designed to see whether a pong ball is capable of transferring bacteria to beer, assuming bacteria are on the ball in the first place. To test this, pong balls were inoculated with a non-pathogenic strain of *E. coli* containing a fluorescent gene by dipping each ball in a solution containing 1,000,000 bacteria/mL. This strain of *E. coli* was used so that only the bacterial cells inoculated on the Ping-Pong balls would be counted in the beer. We did this by looking at the number of bacteria on plate count agar under ultraviolet light. Like all the experiments in this book, this one was conducted in a controlled laboratory setting.

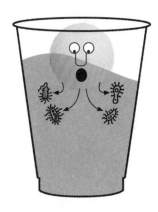

Science Stuff Ahead

Inoculated balls were held with sterile forceps and allowed to drain for 10 seconds and then placed in a plastic cup containing 30 mL of beer. The ball was left in the beer for 10 seconds. Then a sample of the beer was taken to determine the population of *E. coli* transferred from the Ping-Pong ball to the beer. The sampled beer was then serially diluted in 0.1 percent peptone water, surface-plated in duplicate on tryptic soy agar petri dishes, and incubated for 48 hours at 37°C. The dilutions yielding plates with 25–250 colonies were counted, and the population was converted to colony-forming units per ball (CFUs/ball). Control samples were created by aseptically placing uninoculated pong balls in beer. Then the beer was sampled for bacteria using the same procedure as described for inoculated samples. The bacterial populations recovered from the beer exposed to the inoculated balls were compared to the populations recovered from beer exposed to uninoculated balls as well as from beer that had not been exposed to Ping-Pong balls. This experiment was replicated three times with four observations for each treatment. Sample

means (averages) and standard deviations were determined for each treatment using the Statistical Analysis System (SAS).[5]

Results of Experiment 2-2

Drink, drink, drink? In our controlled tests, nearly all *E. coli* bacteria inoculated on a ball were transferred to the beer within the 10 seconds the ball was left in the beer. An *E. coli* population averaging around 530,000 cells/mL of beer was detected after exposure to pre-inoculated pong balls, but no *E. coli* cells were detected in beer after exposure to non-inoculated pong balls. Furthermore, no *E. coli* cells were recovered from the beer not exposed to Ping-Pong balls at all (controls) (Figure 2).

In conclusion, beer pong balls do collect environmental bacteria from

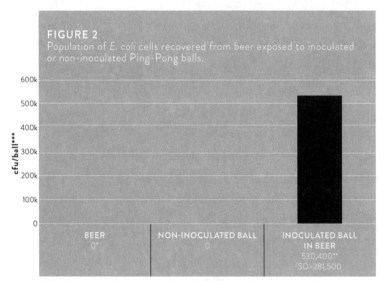

FIGURE 2
Population of *E. coli* cells recovered from beer exposed to inoculated or non-inoculated Ping-Pong balls.

*No cells were recovered at the detection limit of <10 cells; n=12
**Mean with standard deviation (SD).
***CFU/ball = colony forming units (bacteria) per Ping-Pong ball.

contaminated hands and surfaces. And pong balls do transmit bacteria to beer. So if you don't think playing beer pong can spread disease, think again. Back in 2009, the Rensselaer Polytechnic Institute, a university in upstate New York, banned beer pong games due to concerns that the game had caused an unusually high number of cases of bird flu in college students. Furthermore, Billy Gaines, owner of BPONG, a company that organizes national beer pong championships, said that pong contestants had complained of coming down with "pong flu" at pong tournaments. Mr. Gaines was not sure if the malaise was due to spreading of germs during the game or to excessive consumption of the fermented elixir.

Pong flu could possibly be some combination of both factors, but here's something else to consider: The longer a beer pong game lasts, the tipsier the players become, and the more they need to relieve themselves. And the less likely they are to use good hygiene practices when visiting the lavatory before returning to the game, perhaps to throw a ball in your cup of beer with their unwashed hands. Now that you know about pathogens and bacteria transmission, isn't that a nice thought to contemplate?

THINGS TO CONSIDER

Is it wise to participate in games like beer pong? The answer depends on whether you are willing to risk being contaminated to play the game. Keep in mind that the larger the population of bacterial contaminants on the pong ball surface, the greater the risk of transferring microorganisms to the beer. Other factors to consider in this debate include the types of microorganisms present (that is, human pathogens or harmless microorganisms), the human infectious dose of the pathogen required to contract the disease, the general health status of the game participants (healthy versus immunocompromised), and whether the ethanol concentration in the beer and contact time with the ball are sufficient for the alcohol to kill any bacteria present. Ethanol has been shown to exert an antimicrobial effect on some microorganisms; however, the typical alcohol (ethanol) concentration of beer is too low to have much effect on microorganisms.[6]

Essentially, the type of organism present and the general health status of the participant are what determines the minimum population of bacteria/viruses required to make someone sick. These are factors to consider in determining the true risk associated with playing beer pong. Other things that affect bacterial transfer include surface properties (roughness, moistness), pressure applied during contact, contact time, and biofilms, as mentioned in the introduction to Part 1 of this book.[7] We know there are differences in how bacteria adhere and transfer from these surfaces, but don't start thinking it's safer to play beer pong on a vinyl floor rather than outdoors. You are, after all, at the mercy of what gets on the ball and how sick it could make you. So "don't hate the game"—just understand that where the ball goes can create a problem.

3

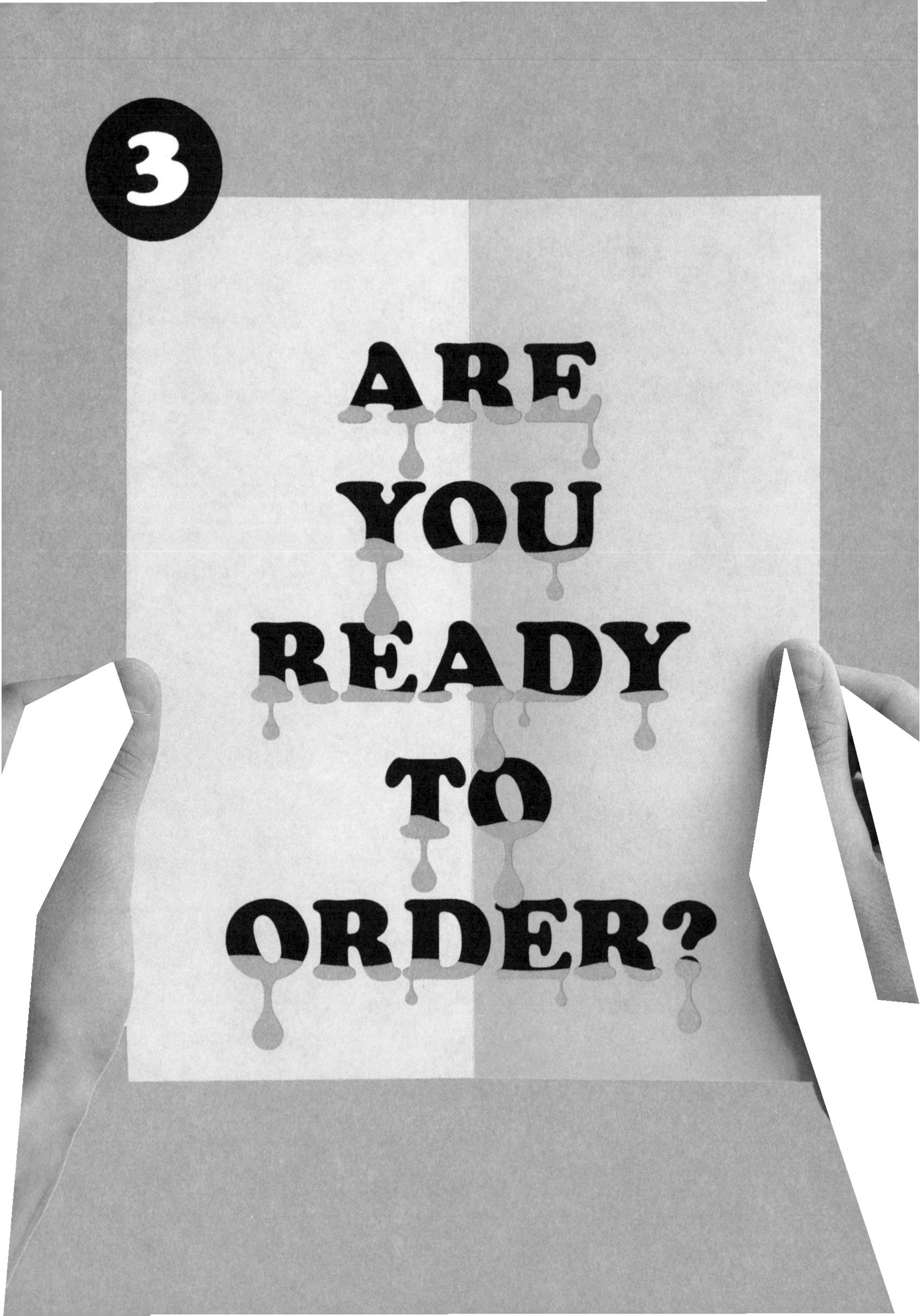

Chapter 3

ARE YOU READY TO ORDER?

Who's been touching that menu you are holding? Well, just about everyone. Have you ever thought about who touches your menu before you place your order? Restaurant menus are handled by nearly everyone who walks into and works at a restaurant, and yet most people give little thought to how clean menus are or who touches them. Bacteria are very likely transferred from one diner to another through handling of menus. It also might matter what type of material the menu is made of, since apparently bacteria transfer faster from laminated menus than paper ones.[1]

How often do Americans eat out? The answer is a whole lot. In 2015, the National Restaurant Association reported $783 billion in restaurant sales at over one million locations in the United States.[2] About four out of ten of us eat out on a daily basis. For 2016, that number translates to 19 million people visiting full-service, sit-down restaurants and 46.5 million dining at fast-food restaurants.[3] Because all of that eating out requires lots of people to make and serve the food, U.S. food industry employees account for 19 percent of the nation's workforce income.

Can you get sick from eating out? Absolutely. Of the 19,119 documented foodborne illness outbreaks reported to the Centers for Disease Control and Prevention between 1998 and 2015, a single location where food preparation takes place was identified as the source for 13,857 cases—that's 72 percent of the outbreaks. The CDC reports that 63 percent (11,992 cases) of the outbreaks were linked to food prepared at a restaurant or deli while 10 percent (1,916 cases) were associated with food prepared by a catering or banquet facility (Figure 1).[4]

Using epidemiological data from North America, Europe, Australia, and New Zealand, Elizabeth Redmond and Christopher Griffith from the University of Wales in Cardiff reported that only 12 percent (2,314 cases) of the outbreaks were related to food prepared in private homes. It's no surprise that public dining places such as restaurants, cafeterias, and bars are locations most often cited for foodborne illnesses and food-related diseases.[5] These types of establishments were responsible for 54 percent of outbreaks in the UK between 1993 and 1998, and they were associated with 45 percent of outbreaks in the United States.[6] It is important to note that most foodborne illnesses go unreported. The CDC estimates that 48 million cases of food poisoning occur annually in the United States and, as was mentioned in the introduction, that one in six Americans gets sick from eating tainted food each year. These estimates are partially based on the 19,119 documented outbreaks of foodborne illness from 1998 to 2015 cited in the previous paragraph, of which there were 373,531 cases of illness, 14,681 hospitalizations, and 337 deaths. Case-control studies have reported that people who often eat out are

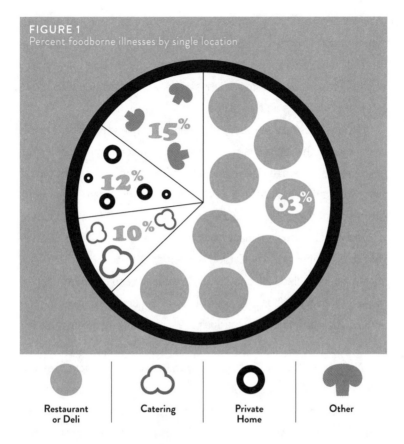

FIGURE 1
Percent foodborne illnesses by single location

15%

12%

10%

63%

Restaurant or Deli | Catering | Private Home | Other

more likely to get food poisoning than those who eat at home.[7] And if you don't think those health rating signs posted at restaurants mean anything, you'd better think again. In fact, more people get sick at restaurants with low food safety ratings compared to those rated high.[8]

By now, we know full well that bacteria and viruses reside on hands and surfaces, and the presence of microorganisms can indicate poor sanitation. For example, repetitive hand contact with contaminated surfaces increases the spread of disease from person to person. This most assuredly includes restaurant staff.[9] Maybe we need more than just signs to remind people to wash their hands after visiting the restroom. In November 2003, one foodborne disease outbreak caused 601 patrons to contract hepatitis A from a single Pennsylvania restaurant and resulted in 124 hospitalizations and three deaths.[10]

So, what's being done about this? Personal hygiene practices are the characteristic that customers most often use in defining food safety at restaurants.[11] Due to the high percentage of foodborne illness associated with eating out, public health officials use routine restaurant inspections to prevent foodborne illness and promote safe food-handling practices.[12] In California, Los Angeles County devotes approximately $10 million each year to these inspections. Effective interventions to decrease cross-contamination include adequate hygiene of hands and food contact surfaces.[13] When a foodborne disease outbreak occurs at a restaurant, the owners may face financial losses due to serving fewer customers and dealing with lawsuits[14] that might lead to fines, bankruptcy, or even jail time.

And how are these efforts going, you might ask? Even though restaurants in the United States undergo local health department inspections, we found a report showing that many of these restaurants (60 percent) regularly have subpar food hygiene practices.[15] Two criteria most closely related to reducing the risk of foodborne illness in restaurants were (1) the person in charge demonstrating knowledge of preventing food-related illness and (2) preventing cross-contamination.[16]

MENUS AND MICROBES

So, there you are, sitting down at a restaurant with your family to order a meal. We don't want to spoil your meal, but we know that nearly all public surfaces have bacteria on them, so you are likely surrounded by them. While it's true that menus are often laminated with plastic to prolong their use and protect them from food and drink, we also know that plastic surfaces can certainly harbor bacteria and were found to transfer more bacteria than paper menus. Environmental conditions also affect bacterial populations on menus and can therefore impact consumers. For example, *Listeria monocytogenes*, among other foodborne pathogens, can adhere to plastic surfaces and potentially be transferred from the menus to hands.[17] So, the menu you are holding has probably had something spilled on it at some point, has definitely been touched by strangers, and is likely crawling with bacteria.

Knowing this, we feel that it is important to examine the possible role of menus as a vehicle for harboring and transmitting bacteria. So we posed a question that has two parts: (1) What is the average number of bacteria on menus, and (2) how many bacteria are typically transferred from a dirty menu to hands, and from dirty hands to a menu?

Before starting our own experiments, we wanted to see what other studies have found about menus and bacteria. *Salmonella* species (spp.) and *E. coli* can survive for up to 72 hours on laminated plastic menus, but these bacteria lasted only 6 hours on paper menus (at that point they fell below detection levels of 100 cells/cm^2).[18] This means that bacteria can be transferred from wet, laminated menus for up to 72 hours after contamination, and up to 6 hours after inoculating dry paper menus. It turns out that bacteria can persist longer on laminated menus than on paper ones.

So let's see how much bacteria really get on—and stay on—those menus. We are going back to the laboratory to look at (1) menus in local restaurants that were randomly sampled for bacteria (total aerobic bacteria and *Staphylococcus* spp.); (2) the transfer of bacteria from hands to menus; and (3) the transfer of bacteria from menus back to hands.

As a side investigation, we also wondered what parts of the menu get handled the most. We conducted tests to determine which areas on a menu the consumers most often touch. A luminous cream (Glo Germ gel from Science Bob Store) was spread on the hands of six participants before they handled menus. The menus were then examined under a UV light (8-watt handheld model made by UVP) to see where the menus showed the most signs of being touched. In a previous study conducted at Purdue University, the researchers said that consumers most often touch the outside edges of the menu.[19] We found this to be true, but we also discovered that people sometimes touch the middle of the menu when choosing a meal.

EXPERIMENT 3-1: Sticky Business

Materials and Methods

Our microbiology students fanned out all over Clemson with sterile swabs specifically designed to sample surfaces such as menus. Eighteen different restaurants in the area (divided into six categories) were sampled, for a total of 216 samples collected over an eight-month period. The sampling times were also categorized as periods of high traffic (lunch from 11:30 a.m. to 2:00 p.m. and dinner from 5:00 p.m. to 8:00 p.m.) and low traffic (all other times). To sample the menus, swabs were taken from a capped tube containing a sterile solution, put back into the capped tube, and returned to the lab in a cooler for further testing.

Science Stuff Ahead

Swab-Samplers (made by 3M Swabs, 3M Company) were used for menu sampling time because of their ease of use, affordability, and rapid sampling time of environmental surfaces.[20] Swabs were taken to restaurants in a cooler bag (Everest cooler bag) and kept cool at all times until plating was completed. The menu-swabbing technique used a zigzag pattern for a total of five lines from left to right, from top to bottom, from the top left corner to the bottom right corner, and from the top right corner to the bottom left corner (for a total of 20 lines). The restaurant menus sampled fell into three general sizes of around 603, 768, and 1,207 cm^2. The average linear distance covered by the 20 swab lines was 57 cm.

Back in the laboratory, sample tubes containing the swab and sterile 0.1 percent peptone water were vigorously shaken by hand for 10 seconds under a biosafety hood (Labconco Purifier 36208-02 Class II/A Laminar Flow Biohazard Hood, made by LABEQUIP LTD) to release bacteria from the swab into the sterile diluent.

Staphylococcus species and total plate count (TPC) Petrifilm plates (3M Company) were used to enumerate aerobic bacterial populations on menus. Petrifilm plates were placed in a 37°C incubator (VWR sympHony

Gravity Convection Incubator) for 24 hours for the *Staphylococcus* spp. Petrifilm and 48 hours for the TPC Petrifilm. After incubation, bacteria were counted using a colony counter (Quebec Darkfield Manual Colony Counter, Reichert Technologies). Bacterial populations were reported as colony-forming units per 15-cm^2 sampling area on a menu (CFUs/15-cm^2 sampling area). Six different menus (two per visit) were sampled at 18 restaurants during periods of high and low traffic. Restaurants were grouped into the following types for analysis: Mexican (4 each), bar (3), pizza (2), steakhouse (2), upscale (4), and other (3). Data were analyzed using the Statistical Analysis System (SAS).[21] Main effects (high/low traffic, restaurant type, replication, day of week) and interactions were tested to determine statistical differences at a significance level of 5 percent.

Results of Experiment 3-1

The number of bacteria recovered from restaurant menus varied considerably between restaurant types as well as between busy and slow times.

Just like total bacterial counts, *Staphylococcus* populations also represent the number recovered from a 15-cm^2 area and for those where "zero" bacteria are shown; the actual populations recovered were below the detection limit of 10 cells (Figure 2). When we extrapolated to the

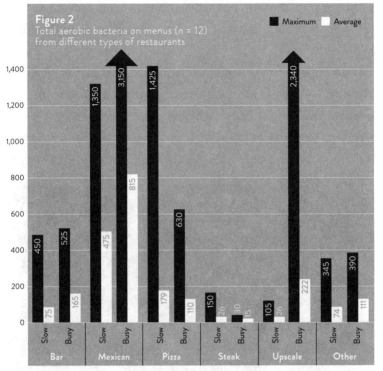

Figure 2
Total aerobic bacteria on menus (n = 12) from different types of restaurants

■ Maximum □ Average

The bacterial populations were expressed as the number of aerobic bacteria recovered from the 15 cm² area of the menus sampled. If this number is extrapolated to the total area of the menus sampled (603, 768, and 1,207 cm²), the total menu bacterial population is estimated to be 40 to 80 times greater than the values shown here.

total area of each menu sampled, aerobic bacteria populations for the 603; 768; and 1,207 cm² menu sizes were 6,030; 7,680; and 12,070 total population, respectively. Generally, more bacteria were detected during busy times than slow business times, and the Mexican restaurant menus yielded higher bacterial populations. Extrapolated *Staphylococcus* populations for the total area of each menu were as follows: 1,648 (603 cm²), 2,100 (768 cm²), and 3,300 (1,207 cm²); Figure 3. Extrapolating these populations is probably an overestimation, but it gives some idea of the potential number of bacteria on the menus, assuming a uniform distribution of bacteria.

The number of restaurants sampled does not necessarily represent a large enough sample to draw any conclusive findings about menu contamination as related to the type of restaurant. However, this snapshot does clearly indicate that menus are contaminated with bacteria including *Staphylococcus* spp (staph). Just to give you a heads-up on staph:

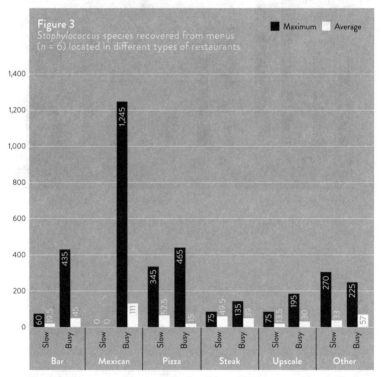

Figure 3
Staphylococcus species recovered from menus (n = 6) located in different types of restaurants

The populations were expressed as the number of staphylococcal bacteria recovered from the 15 cm² area of the menus sampled. If this number is extrapolated to the total area of the menus sampled (603, 768, and 1,207 cm²), the total menu staphylococcal populations is estimated to be 40 to 80 times greater than the values shown here.

this organism is closely associated with human skin and various types of infections, including some relating to food (more on that later).

EXPERIMENT 3-2: Can I Have Some *E. coli* with That Order?

Materials and Methods

Sterile test menus were inoculated with a fluorescent *E. coli* strain and then handled by participants for 1 minute, as if they were ordering a meal at a restaurant. Subsequently, any transferred bacteria were recovered from their hands by thorough rinsing for 30 seconds with 40 mL of sterile water. The population of fluorescing *E. coli* in the rinse water was then determined using standard microbiological enumeration procedures. The number of *E. coli* remaining on the menu after handling was determined in the same manner.

Science Stuff Ahead

Plastic laminated menus created from 12.7 cm × 20.32 cm index cards were inoculated with the fluorescent *E. coli* strain by submerging the menus in 160 mL of a bacterial inoculation culture containing between 100,000 and 1,000,000 cells/mL. The inoculum was prepared from a resuspended bacterial pellet from the fluorescent *E. coli* culture. Menus were handled as described earlier. Then bacteria from the menus and hand-rinsing solutions were enumerated using serial dilutions in 0.1 percent peptone water that were plated on tryptic soy agar and incubated at 37°C for 24 hours.

The following day the plates were inspected under UV light, and petri dishes from dilutions having 25 to 250 CFUs/plate were chosen for counting. The number of colonies per plate was multiplied by the appropriate dilution factor used in the plating process to arrive at the bacterial population per menu. Control menus that were not handled by any participants were also tested to verify that none of the fluorescent *E. coli* had contaminated the menus. Plates were examined under UV light and only fluorescent bacterial colonies were counted. Eight people participated in the study over a three-day period. The mean and standard deviation based on gender and the predominant hand (right-handedness or left-handedness) were calculated using SAS.[22] T-tests were also conducted to determine if the average transfer to the right hand (log CFU Right Hand) differed from transfer to the left hand and for overall bacterial transfer based on left- versus right-handedness.

Results of Experiment 3-2

Approximately 11 percent, or an average of 3 million, of the bacterial cells on inoculated menus were transferred to both hands during the handling of menus (Figure 4). Again, the percentage of bacteria transferred between participants ranged widely, from 32 percent to less than 1 percent (Figure 4.1). Overall, the right hand acquired more bacteria from the menu than the left hand, due partially to having more right-handed participants and therefore a dominant hand effect.

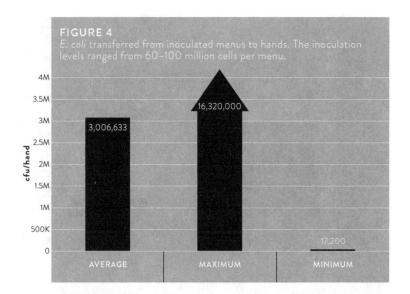

FIGURE 4
E. coli transferred from inoculated menus to hands. The inoculation levels ranged from 60–100 million cells per menu.

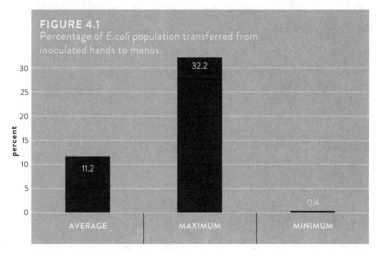

FIGURE 4.1
Percentage of *E.coli* population transferred from inoculated hands to menus.

Lefty versus Righty When handedness (left or right) was considered, the dominant hand tended to acquire more bacteria from the inocu-

FIGURE 5
Percent of *E. coli* transferred due to handedness

right-handed   left-handed

a–c Means with different letters are significantly different at a level of 5%; n=12 for lefties and 36 for righties.

lated menu. Right-handed participants also showed a greater differ-ence between the number of bacteria picked up from handling the menu between the dominant and nondominant hand when compared to left-handed participants (Figure 5).

EXPERIMENT 3-3: Bacteria Hanging Out on Menus

Materials and Methods

To determine how many bacteria survive on menus, we inoculated them with our fluorescent *E. coli* strain and then counted the survivors after 24 and 48 hours.

Science Stuff Ahead

Nine menus per three replications (for a total of twenty-seven observations) were inoculated with 0.1 mL of a 5–6 log population of *Escherichia coli* JM109 that was prepared as described in Experiment 3-2. Menus were held under ambient conditions (~78°F, ~35 percent RH) and surviving bacteria recovered by rinsing with 40 mL of 0.1% peptone water. Initial *E. coli* populations on menus were determined after 30 minutes as well as after 24 and 48 hours of drying at ambient temperatures and relative humidities. The menu rinse recovery solution was serially diluted and surface plated, also as described in Experiment 3-2. Plates were counted after 24 hours at 37°C as described previously. Main effects—replication and holding time—and their interactions were tested to determine statistical difference at a significance level of 5 percent using SAS.[23] The survival rate was calculated according to the equation $N/N_0 \times 100 =$ percent survival rate, where N_0 is the number of CFUs/mL at zero time, and N is

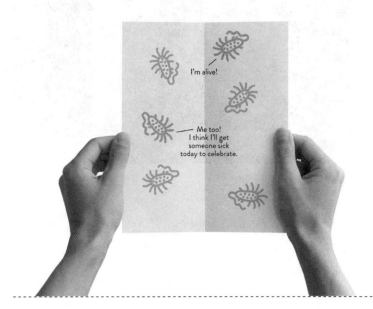

the number of CFUs/mL in the samples after they had been kept at room temperature for 24 or 48 hours.

Results of Experiment 3-3

While the percentage of bacteria surviving on the menus after 24 and 48 hours was low (~1–2 percent), survivors were still around 200,000 bacteria and there was very little change in the number of surviving bacteria between 24 and 48 hours (Figure 6). A highly contaminated menu can thus remain "hot" for days after the initial contamination if not cleaned and sanitized.

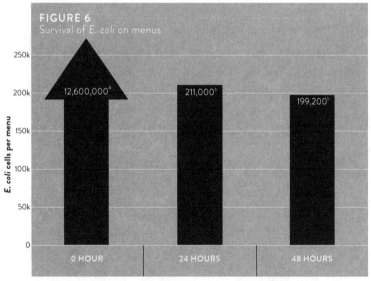

FIGURE 6
Survival of *E. coli* on menus

[a,b] Means with different letters are significantly different at a significance level of 5%.

THINGS TO CONSIDER

E. coli on Menus   *Escherichia coli* can adhere to plastic without a significant loss in population after 48 hours and only decreases from 100 million to 10 million cells after 6 days.[24] Other human pathogens such as *Listeria monocytogenes, Staphylococcus aureus*, and the hepatitis virus can also adhere to plastic and may be present on menus. We have reached three important conclusions:

1. Bacteria are present on menus.
2. They can survive for long periods of time.
3. They can be transferred to the hands of consumers by touching the menus.

Therefore, regular cleaning and sanitizing of menus may be a good preventive practice for reducing the risk of foodborne disease and other illnesses.

Staph on Menus   *Staphylococcus aureus* is a pathogen that is commonly associated with human skin, produces a heat-stable toxin, and is in some cases highly antibiotic resistant. This bacterium, generally spread by human contact, causes both foodborne and non-food-transmitted illnesses such as skin infections (boils, methicillin-resistant *Staphylococcus aureus* [MRSA] infections, and so on). Heat-resistant staphylococcal enterotoxins are produced by some strains of staphylococcus that cause staphylococcal food poisonings. Staph can be routinely found in nostrils, on skin, and in hair. Approximately 30–50 percent of the human population carries *Staphylococcus* species, and staph can live and thrive at temperatures from 45°F to 119°F, at pH values from

4.2 to 9.3, and at salt concentrations of up to 15 percent.[25] This wide range of growth parameters enables *Staphylococcus* species to survive and grow in many different places and environments.[26] Data from the present study identified staph on plastic-laminated restaurant menus. Research at the Shriners Hospital in Cincinnati by Alice Neely and Matthew Maley found *Staphylococcus aureus* was capable of adhering to plastic and also surviving on plastic for at least one day, and that some staph species were capable of surviving up to 56 days on certain plastic materials such as polyester and from 22 to 90 days on polyethylene.[27]

MORE THINGS TO CONSIDER

We've just determined that restaurant menus harbor bacteria and thus may be a health risk for you as the diner. We can, of course, see how menus get dirty: They are handled by hosts, servers, and workers busing tables. And when customers come into the restaurant, whatever bacteria are on their hands can then end up on the menu. We looked at other studies from outside of our lab. While conducting her independent study of twelve restaurants in Abilene, Texas, high school student Michelle Bridgestock found *E. coli*, staph, and bacteria from soil on menus.[28] Additionally, a 2014 press release covered studies showing that menus can harbor more bacteria than toilet seats.[29] We get it, that's probably not what you want to hear. But it's true! Moreover, restaurants that fail to clean and sanitize their menus regularly also contribute to the risk of cross-contaminating their restaurant patrons. So, what to do? Well, we recommend that you take matters into your own hands—wash and sanitize them *after* ordering your food.

Clean
Me!

part 2

Air &

In this part of the book, we take a look at how air and water can dole out microbes. This happens when you do things like blow out birthday candles, improperly wash your hands, and use electric hand dryers. One obvious way that humans emit bacteria and other microbes into the air is by simply breathing, but we also do things like speaking, coughing, sneezing, and laughing. Of course, we might think of these microbes as airborne bacteria—which they are—but the bacteria are actually riding on tiny water droplets like little paratroopers on water balloons.

In Chapter 4 of this part, we consider numerous studies that have examined the number, size, and contents of droplets emitted by humans into the air from their mouths. Over two thousand moisture particles are released per breath, all less than 5 microns (μm) in diameter.[1] (For reference, a human hair is between 50 and 100 µm in diameter, and dust mite feces are 10 µm in diameter. We can't miss a chance for a feces analogy, now can we?)

The average bioaerosol particle released by coughing and speaking is 13.5 µm for coughing and 16.0 µm in diameter for speaking.[2] Bacteria are between 0.5 and 5.0 µm in length while viruses are much smaller, at around 0.02 to 0.3 µm in length. Thus water droplets in human breath are large enough for bacteria as well as viruses to ride on—which is just great, isn't it? *Staphylococcus* species (spp.) as well as *Streptococcus, Corynebacterium, Haemophilus*, and *Neisseria* species are commonly found in respiratory bioaerosols, and pathogens such as *Streptococcus pneumonia* and *Staphylococcus aureus* are also found in human breath.[3] Several researchers have concluded that airborne transmission is a likely pathway for contagious diseases like the flu.[4] Viral influenza also was found in the exhaled breath of infected patients. In one study, 60 percent of patients tested had detectable levels of the virus in exhaled breath; in another study, 81 percent of flu patients had influenza in their breath.[5] Twenty-five percent of tuberculosis patients also released between 3 and 633 TB cells (*Mycobacterium tuberculosis*) per cough, verifying that expelled bioaerosols carry both bacteria and viruses.[6]

In Chapter 5, we take a look at hand washing: Does how we wash and dry our hands affect how clean we get them? What is the best hand-washing procedure? Well, the U.S. Food and Drug Administration (FDA) Food Code has a five-step method that is also endorsed by the National Restaurant Association's ServSafe training program. The FDA Food Code is only a set of recommendations, since each state has the authority to set laws and regulations for food retailers.

Many states require ServSafe training and use a version of the hand-washing procedure found in the Food Code. The FDA Food Code specifies this procedure:

1. Rinse hands under warm running water.
2. Apply soap (cleaning compound) in the amount recommended by the soap maker.
3. Rub hands together for 20 seconds (originally 10–15 seconds).
4. Rinse hands under clean, running water (minimum of 10 seconds).
5. Dry hands with a paper or cloth towel or by using an electric hand dryer.

Hand-washing practices affect food safety in many environments: hospitals,[7] food-service facilities,[8] food processing establishments,[9] homes,[10] fairs,[11] and childcare centers.[12] A study by Rebecca Montville and her collaborators reported that washing hands with soap and drying them with a paper towel had the greatest influence on reducing the number of bacteria remaining on hands after washing (that is, removing more than a hundred bacterial cells per hand).[13] Researchers considered other factors that had negligible effects on hand-washing effectiveness. These included faucet type (touch-free, clean spigot, inoculated spigot), type of soap (non-antimicrobial, antimicrobial, soap with chlorhexidine gluconate), and presence of a ring (hand washing with or without a ring on your finger). When it comes to cleaning hands, we know that (1) when used properly, gloves are better than no gloves;[14] (2) sanitizers are better when used after hand washing;[15] and (3) water temperature was not definitive, a finding recently substantiated by Donald Shaffner's group at Rutgers University.[16]

In Chapter 6, you will see that hand drying using either an electric dryer or paper towels has become a hot topic. It might seem environmentally friendly to use the electric dryer, but we'll soon explain why you might be worse off in the area of cleanliness. Before diving into upcoming chapters, we'll leave you with this bit of fun trivia: Research has found that every time a toilet is flushed, water droplets carrying bacteria are dispersed into the air. Think about that the next time you flush.

4

BLOWING OUT

BIRTHDAY

CANDLES, OR

SPRAYING GERMS

ON CAKE?

Chapter 4

BLOWING OUT BIRTHDAY CANDLES, OR SPRAYING GERMS ON CAKE?

Suppose you are attending Aunt Bessie's ninety-first birthday party, and she's ready to make a wish and blow out the candles on her cake. She needs some help with blowing out the ninety-one candles blazing away in front of her, so she enlists her three young great-grandchildren. All together now, the four of them blow, and blow, and blow again. It's, well, a lot of blowing over the cake everyone is about to eat. Do you really want someone blowing on your food just before you eat it? (Or at all, for that matter?) Who started the practice of blowing out candles on birthday cakes?

Several theories are offered to explain when and where the tradition of blowing out birthday candles originated. Some think it began in ancient Greece when people brought cakes with lit candles to the temple of the goddess of the hunt, Artemis. Other ancient peoples believed that smoke from candles carried their wishes to the gods. One of the first documented accounts of using birthday candles appeared in the mid-1700s among the writings of Andrew Frey, who was traveling in Germany at that time. While visiting Count Ludwig von Zinzendorf in Germany, Frey reported that candles (one for each year of the count's age) were placed on cakes for his birthday celebration.[1]

SPRAYING GERMS, MUCH?

Droplets from your respiratory tract and mouth are expelled by coughing, sneezing, talking—and yes, even breathing. In fact, particles of flu virus were detected in the exhaled breath of infected individuals even during normal breathing and talking.[2] When respiratory droplets are released, they may spread infection directly through the air or by landing on a surface.[3] Other studies found that exhaled breath contained from 693 to 6,293 bacteria/cubic meter and that humans pollute indoor air by emitting bacteria at a rate of about 37 million gene copies per person, per hour.[4] So when someone blows out birthday candles, it's extremely likely that their breath carries bacteria or viral particles directly toward the cake that everyone else is about to eat! Let's verify this idea with an experiment.

 EXPERIMENT 4-1: Blow or No Blow

Materials and Methods

The objective of this study was to evaluate the level of bacterial transfer to the top of a cake when someone blows out candles. Since we were concerned only with what was landing on top of the cake, we designed a simulated cake consisting of a thin layer of icing spread on a sheet of foil and cut to fit over a circular foam base (Figure 1). Sixteen candleholders and candles were inserted through the icing/foil and into the foam base.

Each test participant was first asked to smell and consume a piece of hot pizza to generate saliva and simulate a birthday party atmosphere. After the candles were lit, test participants blew until all of the candles on the mock cake (see Figure 1 again) were extinguished. For each testing session, a control sample was collected: the same procedure was followed for the test sample, but candles were not blown out. After the candles were lit, the test participants either blew them out or did not blow them out. The candles and holders were then removed from the foam base without touching the icing. Using sterile forceps, the foil was folded in half with the layer of icing inside.

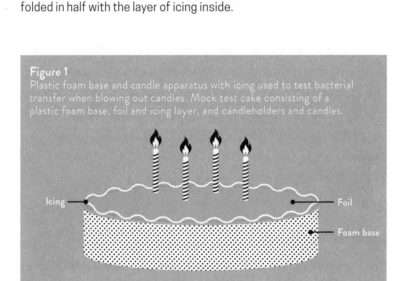

Figure 1
Plastic foam base and candle apparatus with icing used to test bacterial transfer when blowing out candles. Mock test cake consisting of a plastic foam base, foil and icing layer, and candleholders and candles.

Icing — Foil — Foam base

Science Stuff Ahead

The foil with icing was placed in a Stomacher bag (Classic 400, Seward UK), unfolded inside the bag, and then 50 mL of 0.1 percent sterile peptone water was poured into the Stomacher bag over the iced surface of the foil. The Stomacher bag was placed in a Stomacher (Stomacher 400, Seward) and mixed at 230 revolutions per minute (rpm) for 1 minute. A Stomacher is simply a laboratory tool that mixes samples within a bag that simulates the mixing action of your stomach; two paddles alternately squeeze a sealed bag in a stomach-like mixing action. Duplicate 1-mL and 0.1-mL volumes of the homogenate were aseptically removed from the Stomacher bag, serially diluted in 0.1 percent peptone water, and surface-plated in duplicate onto petri dishes containing plate count agar (Difco Plate Count Agar). Plates were incubated at 37°C for 48 hours. Colony-forming units (CFUs) were counted on plates containing 25–250 colonies and then converted to CFUs per icing sample and $\log_{10}$ CFUs per sample.

The experiment was replicated three times on separate days, using 11 participants and yielding 33 observations per treatment (blow or no blow). The effect of blowing versus not blowing out candles on bacterial counts in the frosting was determined using the proc univariate command of the Statistical Analysis System (SAS). Statistical differences between the blowing and not-blowing treatment groups were determined at a significance level of 5 percent. Descriptive statistics such as the mean (average), median, range, and standard deviation were also obtained.[5]

Results of Experiment 4-1

Some studies on airborne droplets originating from human mouths are found as early as 1899, and others were published before the mid-twentieth century.[6] These early studies yielded varying results, but they all agreed that droplets were released into the air by breathing, coughing, and sneezing. Amazingly, 90 percent of the bacteria-carrying droplets remained airborne for 30 minutes, and some smaller droplets hung around for up to 30 hours.[7] In our study, blowing out candles over icing resulted in a population of bacteria recovered from icing that was *fifteen times* higher than the bacterial population from icing where the candles remained lit (Figure 2). Also, the range (difference between the highest and lowest counts) in number of bacteria

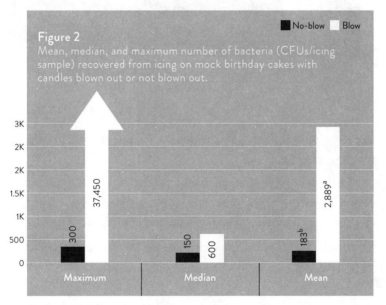

Figure 2
Mean, median, and maximum number of bacteria (CFUs/icing sample) recovered from icing on mock birthday cakes with candles blown out or not blown out.

No-blow = cake icing not exposed to blowing out candles.
Blow = cake icing exposed to blowing out candles.
CFUs/sample = colony-forming units per cake icing sample.
a,b Means within a statistical category with different letters are different at a significance level of 1%; n = 33 observations per treatment.

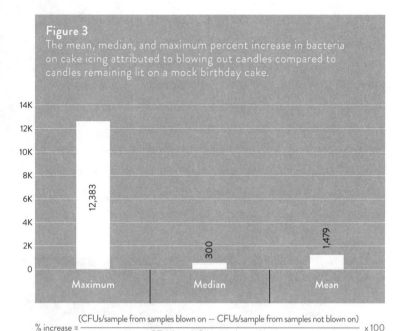

Figure 3
The mean, median, and maximum percent increase in bacteria on cake icing attributed to blowing out candles compared to candles remaining lit on a mock birthday cake.

$$\% \text{ increase} = \frac{(\text{CFUs/sample from samples blown on} - \text{CFUs/sample from samples not blown on})}{\text{CFUs/sample from samples not blown on}} \times 100$$

recovered from icing was 100 times greater for icing exposed to the blow versus no-blow treatment.

Furthermore, the average number of bacteria on the cake icing increased by about 2,700 bacteria, or 1.48 percent, due to candles being blown out (Figure 3). The sloppiest candle blower we tested increased the number of bacteria on the icing by over 37,000, or over a whopping 12,000 percent, compared to the icing not blown on.

THINGS TO CONSIDER

Bacteria are an unavoidable part of life. As nearly everyone knows by now, they are present in and on almost everything people touch or contact. Therefore, we believe it's important to understand how micro-organisms move in our environment and to become familiar with ways to minimize contamination. Disease-causing microorganisms can spread rapidly through a human population by means of bioaerosols, and poor air hygiene can have adverse effects on human health.[8]

Ceremoniously blowing out the candles on a cake might be a problem if the person doing the blowing is sick, since we know that respiratory droplets expelled by coughing and sneezing can carry pathogenic bacteria and viruses.[9] The spread of diseases including SARS and H1N1 avian influenza have been attributed to human bio-aerosols.[10] What's more, microbes originating from respiratory droplets can be transmitted directly through the air or indirectly as droplets landing on surfaces that are then touched (or eaten, in the case of birthday cake). Now that you know the truth about blowing out birthday candles, do you think the tradition is safe in all cases? Well, you are now armed with the facts, so the decision is entirely up to you. Blow wisely.

5

KEEP YOUR
DIRTY
HANDS
TO YOURSELF

Chapter 5

- - - - - - - - - - - - - - - - - -

KEEP YOUR DIRTY
HANDS TO YOURSELF

The next time you shake hands with someone, you might want to consider this: There's a one in five chance that the person did not wash his or her hands after visiting the restroom.[1] Of those who do wash their hands, 35 percent of men and 15 percent of women don't use soap.[2]

New introductions aside, you might want to consider this more important question: who's actually handling your food? This topic has been extensively studied, as we discovered in a keyword search for *hand washing and food* (using the Google Scholar search engine) that yielded 612,000 related results. According to statistics from the U.S. Department of Labor, there

were 4,438,100 food and beverage servers and related workers in the United States in 2012.[3] Only 6.2 percent of these workers had a college diploma, and it looks as if the number of U.S. workers with either a college diploma or even a high school education is projected to decrease over the next fifteen years.

We've found studies showing that levels of education may affect food workers' understanding of the importance of good food-handling practices.[4] A primary line of defense against any kind of microbial contamination is good hand hygiene, since we know that numerous foodborne disease outbreaks have been linked to hand contamination.[5] In fact, approximately 38 percent of food contamination is related to inadequate hand washing.[6] As part of a food safety certification program using the National Restaurant Association's ServSafe curriculum, 1,448 food workers were surveyed about their hand sanitation practices.[7] In this survey, 79 participants stated that they did not sanitize between tasks, and 78 percent stated they did not use proper hand-washing practices before their training. Furthermore, up to 60 percent of adults did not consistently wash their hands when it was deemed appropriate.[8] Generally, hand-washing techniques were inconsistent and judged unsatisfactory.[9] The ServSafe program promotes a hand-washing procedure that workers in the food industry are sup-

DID YOU JUST EAT THAT?

posed to use. The objectives of our study, then, were to (1) compare the efficacy of the ServSafe hand-washing process to simply rinsing with warm or cold water in Experiment 5-1; (2) examine washing practices for gloved hands in Experiment 5-2; and (3) evaluate the efficacy of alcohol-based hand sanitizers on decontaminating inoculated hands in Experiment 5-3.[10]

EXPERIMENT 5-1: To Wash, Rinse, or Not Wash at All?

No Wash   Cold Rinse   Warm Rinse   "ServSafe" protocol

Materials and Methods

We inoculated ground beef with the ampicillin-resistant strain of *Escherichia coli*[11] that possesses a fluorescent gene, the same substance we used for the menu study in Chapter 3. Before starting the experiment, all participants were instructed to wash their hands using the recommended ServSafe protocol. The protocol calls for hands to be rinsed with warm water (104°F/40°C), lathered with soap (without added sanitizers for our study) for 20 seconds, and then rinsed again with warm water (104°F) for another 10 seconds.

Next, 100 grams of inoculated ground beef (100 g = ~18 quarters) were kneaded with both hands for 30 seconds by each of the three participants, who then air-dried their hands for 15 seconds. The four hand-washing treatments—no wash, cold rinse, warm rinse, and ServSafe protocol—were all tested separately. For the cold rinse treatment, participants rinsed their hands with cool running water (79°F/26°C) for 15 seconds. This same method was also used for the warm rinse treatment using 104°F water. No washing or rinsing was used for the no-wash control treatment. Following each treatment, participants were given commercial (non-sterile) paper towels to gently pat their hands dry. To recover the remaining bacteria from participants, each hand was placed in a pre-labeled stomacher bag containing 50 mL (about 10 teaspoons or 3.5 tablespoons) of 0.1 percent peptone water and gently agitated for 30 seconds. Agitation was achieved by the participants gently flexing their hands and fingers inside the bag.

Science Stuff Ahead

To prepare the ground beef inoculation culture, 0.1 mL of the *E. coli* culture was pipetted into 10 mL of tryptic soy broth under a Germfree Bioflow Chamber. The inoculum was placed in an agitator located inside an incubator maintained at 98.6°F (37°C) for 16 to 18 hours. Subsequently, the inoculum was centrifuged at 2,700 rpm for 15 minutes to concentrate the bacterial cells in the bottom of the test tube; then the clear fluid in the top of the tube was decanted and discarded. Next, 10 mL of 0.1 percent peptone water was added to the cell pellet, and the cells were resuspended using a vortex mixer (Fisherbrand Genie 2). The population of *E. coli* in the suspension was about 10^7 per mL. Then 10 mL of the suspension was added to 100 grams (the mass of 20 nickels) of ground meat, resulting in about 10^6 (1,000,000) *E. coli* cells per gram of meat. We poured the *E. coli* suspension evenly onto the meat surface and distributed it throughout the meat by kneading with sterile gloved hands before anyone handled the meat with their bare hands (Experiment 5-1).

We obtained the ground beef from a state-inspected meat processing facility. Each 100 g of ground beef was used for one of the four hand-washing treatments. The experiment was repeated three times on different days, using three separate meat batches and three different inoculation cultures. Enumeration of bacterial populations in the inoculated meat was determined by placing 11 g of meat into a stomacher bag containing 99 mL (1/2 cup = 118 mL) of 0.1 percent peptone water followed by stomaching for 1 minute at 230 rpm. The food homogenate was then serially diluted (by factors of 10) in 0.1 percent peptone water and surface-plated in duplicate onto tryptic soy agar plates. All plates were then inverted and incubated at 36–38°C for 24 hours, after which colonies were counted (with a Leica Quebec Darkfield Colony Counter).

Populations recovered from the hand rinses were determined using the same enumeration method as described above. Bacterial populations were estimated by counting each visible colony that appeared fluorescent under UV light. The results were reported as colony-forming units (CFUs) per milliliter of the diluted beef homogenate or hand-rinsing solution (CFUs/mL of rinse or on a per hand basis). The number of bacteria recovered from the hands (CFUs/hand) was calculated by multiplying the CFUs/mL times the rinse volume.

Remember that *colony-forming units* is the way microbiologists talk about single bacterial or fungal cells that can divide and produce single colonies. In addition, the percent reduction and percent log reduction in

EXPERIMENTAL DESIGN FLOWCHART

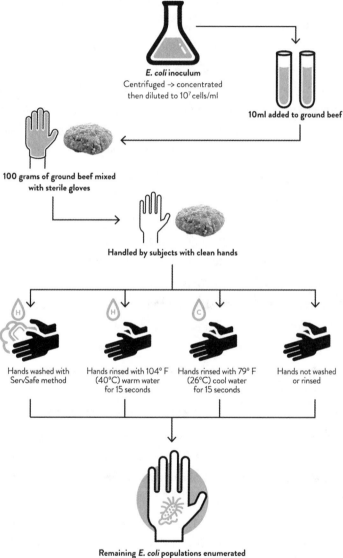

E. coli inoculum
Centrifuged → concentrated
then diluted to 10^7 cells/ml

10ml added to ground beef

100 grams of ground beef mixed
with sterile gloves

Handled by subjects with clean hands

Hands washed with
ServSafe method

Hands rinsed with 104° F
(40°C) warm water
for 15 seconds

Hands rinsed with 79° F
(26°C) cool water
for 15 seconds

Hands not washed
or rinsed

Remaining **E. coli** populations enumerated
from the hands after treatments

bacterial populations per hand/glove of rinse (% log CFUs/hand and % CFUs/hand reduction) were calculated using the following equations:

$$\% \text{ hand reduction} = \frac{\text{no wash (CFUs/hand)} - \text{wash treatment}^* \text{(CFUs/hand)}}{\text{no wash (CFUs/hand)}} \times 100$$

$$\% \text{ log hand reduction} = \frac{\text{no wash (log CFUs/hand)} - \text{wash treatment}^* \text{(log CFUs/hand)}}{\text{no wash (log CFUs/hand)}} \times 100$$

* Wash treatment is either cold rinse, warm rinse, or ServSafe protocol.

EXPERIMENT 5-2: What about Gloves?

Materials and Methods

The glove-washing experiment was conducted identically to the hand-washing treatment described in Experiment 5-1. In Experiment 5-2, however, participants wore gloves during the hand-washing treatments, as well as when mixing the inoculated ground beef. Food preparation gloves from a university food-service facility were used in this experiment (GlovePlus Latex Free Industrial Vinyl by Ammex).

Statistical Analysis

Each experiment was replicated three times, and each experiment was analyzed separately using the Statistical Analysis System (SAS). Experiments 5-1 (bare hand evaluation) and 5-2 (gloved hand evaluation) were subjected to an analysis of variance to determine if there were any statistical differences in mean *E. coli* populations on hands among the washing treatments using a significance level of 5 percent. Since cleansing treatment effects were significant for each experiment, multiple comparison tests were performed using SAS.[12]

Results of Experiments 5-1 and 5-2

There was a stepwise reduction in bacterial populations remaining on the hands and gloves for the hand-washing techniques tested. From the highest population to the lowest, the order was no wash > cool rinse > warm rinse > ServSafe method (Figure 1). In 2002 Barry Michaels and his coworkers also found an improvement in the removal of bacteria from hands when higher water temperatures (120°F/60°C) were used to wash and rinse hands, whereas greater variation was observed at lower water temperatures (40°F/4.4°C).[13] Unwashed hands averaged 5.0 log CFUs/hand (100,000) while hands washed using the ServSafe method retained < 1 log CFUs/hand (< 10).

For the glove experiment, the warm rinse and ServSafe methods did not differ in the level of bacteria removed, while both were superior to the cold rinse and no-wash treatments (see Figure 1). This finding differed from the bare hand-washing experiments, where a significant stepwise difference in the removal of bacteria was observed (ServSafe > warm rinse > cool rinse > no wash). We observed during our experiment that the meat particles tended to cling to the vinyl surface of the glove material and were harder to remove when washing. Generally, a higher population of *E. coli* remained on the gloves after each washing treatment, compared to levels found on hands. Interestingly, even though meat inoculum levels were identical for Experiments 5-1 and 5-2, unwashed gloves retained an average of 125,893 CFUs/hand while unwashed

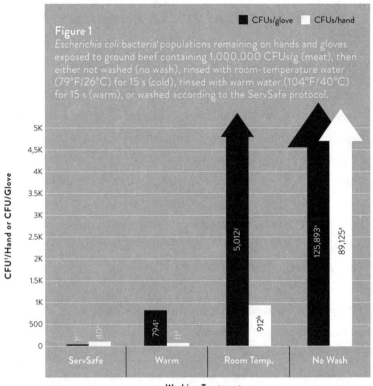

Figure 1
Escherichia coli bacteria[1] populations remaining on hands and gloves exposed to ground beef containing 1,000,000 CFUs/g (meat), then either not washed (no wash), rinsed with room-temperature water (79°F/26°C) for 15 s (cold), rinsed with warm water (104°F/40°C) for 15 s (warm), or washed according to the ServSafe protocol.

Legend: ■ CFUs/glove □ CFUs/hand

Y-axis: CFU[1]/Hand or CFU/Glove

Data labels:
- ServSafe: 1[x], 40[c]
- Warm: 794[z], 11[d]
- Room Temp.: 5,012[y], 912[b]
- No Wash: 125,893[x], 89,125[a]

X-axis: Washing Treatment

a–d Bars with different letters represent means that are significantly different at a level of 5%. Standard error of the mean for log CFU/hand = 0.26.
x–z Bars with different letters represent means that are significantly different at a level of 5%. Standard error of the mean for log CFU/glove = 0.54.
[1] = colony-forming units

hands retained an average of 89,125 CFUs/hand. Thus, based on the bacterial recovery methods used in this experiment, the transfer of *E. coli* was high when meat was kneaded with either bare hands or gloves.

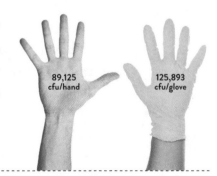

89,125 cfu/hand

125,893 cfu/glove

EXPERIMENT 5-3: Alcohol Sanitizers

Materials and Methods

The sanitizer experiment was similar to Experiment 5-1, except that we applied an *E. coli* inoculum solution directly to hands rather than using meat. A concentrated *E. coli* pellet was added to 850 mL of sterile water and gently mixed. Each of three participants dipped their hands in the *E. coli* solution for 5 seconds and then gently patted their hands dry with a non-sterile paper towel. Hands were then allowed to air-dry for 30 seconds before participants received a dime-sized amount of sanitizer. Hands were rubbed together for 30 seconds and then allowed to air-dry again for 30 seconds. Participants in the no-sanitizer control treatment received no sanitizer, of course, but were otherwise treated the same as participants in the sanitizer treatments. Afterward, participants' hands were separately rinsed for 30 seconds each in stomacher bags containing 50 mL of 0.1 percent peptone water. Four brands of ethanol-based hand sanitizers were tested in the experiment (designated as brand B,

70 percent ethanol; C, 70 percent ethanol; P, 70 percent ethanol; and S, 62 percent ethanol), each containing ethanol and skin conditioners. Experiment 5-3 was replicated three times using three participants each and the

results subjected to an analysis of variance to determine if there were any statistical differences in mean bacterial populations among the sanitizer treatments at a significance level of 5 percent. Since sanitizer treatment effects were significant for each experiment, multiple comparison tests were performed using SAS as similarly described for

Experiments 5-1 and 5-2, except five cleansing treatments were used (no sanitizer and four sanitizers).[14]

Results of Experiment 5-3

We found that sanitizer brand S was not effective in eliminating *E. coli* and did not, in fact, differ from the no-sanitizer control. The most effective sanitizers were brands B, C, and P, although these three were not statistically different from each other, and all three contained 70 percent ethanol. Sanitizers ranked from highest to lowest in cost were P > C >

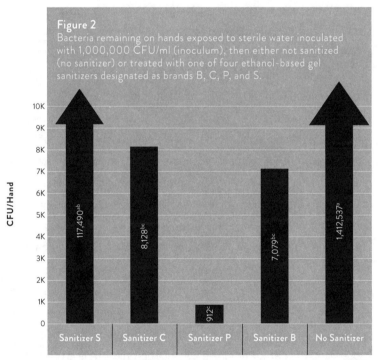

Figure 2
Bacteria remaining on hands exposed to sterile water inoculated with 1,000,000 CFU/ml (inoculum), then either not sanitized (no sanitizer) or treated with one of four ethanol-based gel sanitizers designated as brands B, C, P, and S.

CFU/Hand

Sanitizer S — 117,490^{ab}
Sanitizer C — 8,128^{bc}
Sanitizer P — 912^{c}
Sanitizer B — 7,079^{bc}
No Sanitizer — 1,412,537^{a}

a–c Bars with different letters represent means that are significantly different at a level of 5%. Standard error of the mean for log CFU/hand = 0.65.

B > S. Sanitizer S yielded 1.0 log CFUs fewer *E. coli* per hand (around 1.3 million fewer cells) than the no-sanitizer control, and more than 100,000 CFUs/hand more *E. coli* than the other sanitizers (Figure 2). This finding may be attributed to the lower ethanol content of sanitizer S (62 percent) compared to the other sanitizers (70 percent), or to the addition of other ingredients such as conditioners that may reduce the efficacy of ethanol.

In 2003 Barry Michaels and his colleagues, working out of Michael Doyle's lab, also reported that the combination of hand washing followed by the application of an alcohol-based sanitizer lotion significantly reduced bacterial numbers when compared to hand washing alone, but only when larger quantities (3 and 6 mL) of sanitizer were applied.[15] For example, hand washing followed with 3 or 6 mL of sanitizer resulted in an average reduction in bacteria of over 1,000 CFUs/mL of rinse compared to an average reduction of 100 to 200 CFUs for both hand washing alone or hand washing followed with 1.5 mL of sanitizer.

In this same paper, the Doyle lab group reviewed previously published studies that evaluated other alcohol-based instant hand sanitizers (ethanol, isopropanol, *n*-propanol). These studies reported log population reductions ranging from 0.2 to 5.5. Factors affecting alcohol-based sanitizer effectiveness include alcohol type, alcohol concentration, quantity applied, exposure time, organic load on hands (how dirty), and presence of other ingredients (hand conditioners, dyes/colorants, etc.) that may interfere with the alcohol.[16] The limited effectiveness of alcohol-based hand sanitizers was previously reported by other researchers who applied a 70 percent alcohol solution after washing with a non-medicated soap wash and found no further reduction in total aerobic and mesophilic bacteria. Another study reported no reduction in bacteria after application of alcohol-based sanitizers.[17]

But the news on alcohol-based sanitizers is not all bad. They do reduce the number of stomach ailments, the absences of grade-schoolers who miss class due to illness, and the cases of sickness in university dorms.[18] But a study by the Centers for Disease Control and Prevention (CDC) did support our results and other findings that some alcohol sanitizers are not all they are said to be—specifically, that some commercial sanitizers having lower alcohol content were no more effective than tap water in killing bacteria on hands.[19]

PERCENTAGES AREN'T ALL THEY'RE CRACKED UP TO BE

Reductions in bacterial populations resulting from antibacterial mouthwashes, soaps, and cleansers are often reported as percentages, but they can also be calculated and expressed in different ways. Back in 2003, Brian's lab group contrasted several of these methods while reporting on the transfer of *Salmonella* and *Campylobacter* bacteria between lettuce and food preparation surfaces.[20] Different methods of calculating the percentage of cells removed during hand-cleansing treatments can give drastically different results. For example, expressing the same reductions on a percent CFUs/hand versus a percent log CFUs/hand differed as much as from 94.96 to 28.60, respectively—a difference of over 60 percent (Figure 3). Bacterial populations are often reported in log values since the whole numbers are generally very large and more difficult to express graphically.

Furthermore, expressing these reductions as log numbers reduces

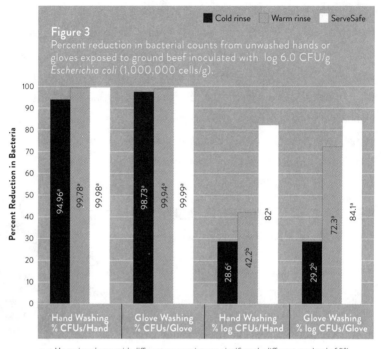

Figure 3

Percent reduction in bacterial counts from unwashed hands or gloves exposed to ground beef inoculated with log 6.0 CFU/g *Escherichia coli* (1,000,000 cells/g).

Legend: ■ Cold rinse ▦ Warm rinse □ ServeSafe

a–c Means in columns with different superscripts are significantly different at a level of 5%.

the influence of extreme values, or outliers, that can skew the data. But when the percentage of change is calculated using log values, the results can be misleading if you don't fully grasp the difference between whole numbers versus expressing the numbers as log values. For example, some companies advertise that their bactericidal and antimicrobial soap products are capable of killing 99.9 percent of any "germs" that may be present. This claim can be confusing for some consumers since people generally don't fully understand what 99.9 percent means when dealing with very large numbers of bacteria. It sounds like nearly 100 percent of the organisms are killed when in fact it means there is a 3 $\log_{10}$ reduction of the population. Bacteria are often present in very high populations on highly contaminated surfaces, so that many bacteria may remain even when 99 percent are killed. For instance, if a contaminated surface had 1 million bacterial cells per square inch (6 $\log_{10}$), which is probably reasonable in some cases, a 99 percent reduction (2 logs) would leave 10,000 cells per square inch (log 4.0). This result was also illustrated in our study: the cold water rinse removed ~95 percent of the bacteria present on the hands, yet over 5,000 bacteria (log 3.7) remained on the hands.

To muck up the discussion even more, manufacturers of hand sanitizers are generally required to test their products against a set group of bacteria, using set concentrations and exposure times, to get a germicide, sanitizer, or disinfectant claim approved. Thus the sanitizers are tested against these specific organisms only and not all organisms that may be present in the real world. Different bacteria—even of the same type—and viruses often differ in their ability to withstand sanitizers. This is especially true for those that biologists refer to as environmental isolates, which are ones that have lived and adapted in the real world.

On a related note, in September 2016, the U.S. Food and Drug Administration (FDA) issued a ruling on the safety of antifungal and antibacterial soaps. It banned nineteen ingredients used in over-the-counter antibacterial soaps; however, the ruling does not include hand sanitizers used in health-care settings. The final rule went into effect in September 2017.[21] The FDA concluded, not without dispute and compelling responses from the Personal Care Products Council, that the manufacturers did not prove that using antibacterial soaps with the listed ingredients are "both safe for long-term daily use and more effective than plain soap and water in preventing illness and the spread of infections."[22] The lack of increased effectiveness compared to soap and water and the

potential risks of long-term exposure leading to the development of bacterial resistance were determining factors in the FDA ruling.[23] Of the nineteen banned compounds, triclosan, phenol, and iodine-containing ingredients are probably most familiar to the general public. Although these compounds are not really toxic to the skin, the FDA thought repeated exposure to these compounds was not safe.

THINGS TO CONSIDER

Several sophisticated techniques have been used to study hand sanitation, including some with interesting names such as quantitative microbial risk assessment (QMRA), Palmar Method, and the Melon Ball Disease Transmission (MBDT) model. These laboratory methods and computer models were developed by researchers and are used by the FDA and personal care products industry to determine the effectiveness of cleansers. As in our study, other researchers also use nonpathogenic bacteria such as *Enterobacter aerogenes* to quantify sanitation practices.

In a study where 1,000,000 *E. aerogenes* cells per square centimeter (cm^2) were inoculated onto chicken skin, 100,000 were transferred to hands. Subsequently, 10^3–10^4 CFUs/cm^2 (1,000–10,000) were transferred from the hands to vegetables upon handling.[24] Percent transfer of *E. aerogenes* between hands, food, and kitchen utensils has been reported as high as 100 percent.[25] Surface areas of the human palm range from 110 to 190 cm^2 for both males and females; median areas are about 120 and 170 cm^2 for females and males, respectively (about the

DID YOU JUST EAT THAT?

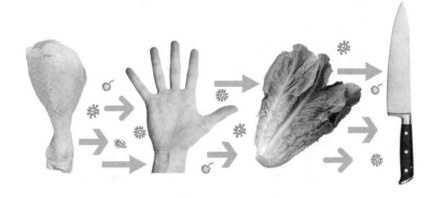

surface area of a Ping-Pong paddle). That's plenty of surface area for harboring high populations of bacteria, which have been estimated between 10,000 and several millions, depending on the level of contamination.[26] Most food industry workers' hands have between 2–3 (100–1,000) and 5–7 log (100,000–10,000,000) CFUs/hand of *Enterobacteriaceae* (a large family of bacterial types including many harmless organisms; this family of bacteria also include several familiar pathogens such as *Salmonella* species and some *Escherichia coli*) and mesophilic bacteria (bacteria that like body temp), respectively.[27] Given what's just been described, we can conclude by saying that proper hand cleansing is vital to preventing foodborne disease.

Improper handling and sanitation practices lead to cross-contamination from person to person and from person to food. These practices ultimately result in 27 percent of reported foodborne outbreaks and infections involving foodborne pathogens.[28] Researchers at the Henry Ford Hospital in Detroit demonstrated that by showing health workers magnified pictures of bacteria recovered from mouse pads, door handles, telephones, and other common surfaces in the hospital, the so-called yuck factor could improve hand hygiene.[29] Compliance with proper hand hygiene reportedly increased 24 percent in health workers after viewing the bacterial images, which the researchers predicted could potentially reduce hospital infection rates by 40 percent.

Going out in public areas and touching handrails have been activities of concern for bacterial transfer even as early as 1900.[30] Riding the public transportation system might be a little like going to a germ convention, since a diversity of pathogens including antibiotic-resistant strains have been found on seats, armrests, handrails, and windows.[31] A trip to the public beach might even be taking a chance, for antibiotic-resistant bacteria have been found in and on sand and shallow water at busy West Coast beaches in the United States.[32] The bottom line is that a good hand-washing regimen is one of the best ways to protect yourself and others from getting sick, although in some cases you might want to make that a good shower instead.

6

HAND DRYERS,

or

BACTERIA BLOWERS?

Chapter 6

- -

HAND DRYERS,
OR BACTERIA BLOWERS?

To dry or not to dry? That is the question. Hand hygiene is a critical factor in reducing the spread of disease-causing microorganisms and has resulted in published guidelines on hand hygiene for health-care workers and retail and food-service workers.[1] Hand drying is an important but frequently overlooked step in adequate hand hygiene. Simply drying your hands can reduce the amount of bacteria remaining on your hands as well as the transfer of bacteria from washed hands to other surfaces by over 90 percent[2] (Figure 1).

Furthermore, while hand drying was found to be an essential step in reducing the spread of methicillin-resistant *Staphylococcus aureus* (MRSA) in hospitals, most hand-washing protocols do not describe a specific hand-drying technique.[3] So, which method of drying hands is best?

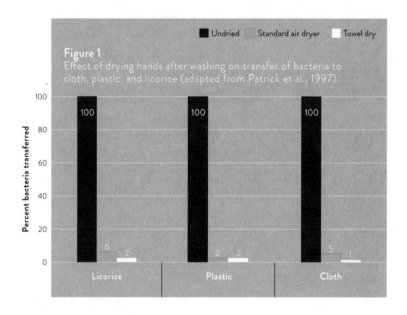

Figure 1
Effect of drying hands after washing on transfer of bacteria to cloth, plastic, and licorice (adapted from Patrick et al., 1997).

Legend: Undried | Standard air dryer | Towel dry

Y-axis: Percent bacteria transferred

Licorice: 100, 6, 2
Plastic: 100, 2, 2
Cloth: 100, 5, 1

TOWEL OR AIR DRY?

According to a study on the environmental impact of using disposable paper towels, air blower hand dryers may be better for the environment.[4] The University of Westminster in the UK—a leader in hand-drying methods for sanitation—has found that air blowers are less effective than towel drying. Their most recent study found that towel-drying spread an average of 1.6 viral plaques over 1 foot from the drying spot while jet-air dryers spread 2,188.7.[5] Another study published by the university reported 35 percent fewer bacteria on hands that were towel-dried.[6] In fact, more disturbing was a study reporting an average increase of 504 percent in total bacteria on fingertips after drying with automatic electric hand dryers, while drying with paper towel and cloth towel resulted in a 42 percent decrease and 10 percent decrease, respectively.[7] The

University of Westminster group also found a 255 percent increase in total bacteria on hands after drying with electric hand dryers, and a 438 percent increase in numbers of intestinal (fecal bacteria) and skin-associated (*Staphylococcus* spp.) bacteria.[8]

It's hard to believe that using an electric hand dryer would increase the number of bacteria on washed hands to the extent found in these last two studies. But the results indicate that electric hand dryers are in fact less effective in removing bacteria from the hands than paper towel drying. And there's more: several studies attributed higher bacterial numbers found on hands after drying from using warm-air and jet-air dryers to the release of bacteria from skin folds and crevices due to rubbing hands together during drying.[9] To make matters worse, using electric hand dryers spreads bacteria from hands to surfaces around the blower by "aerosolizing" bacteria.[10] Despite these findings, FDA Food Code still states that using warm-air dryers is an acceptable method of hand drying for food-service workers.

To see for ourselves, we conducted two experiments to determine how air hand dryers can contribute to the transfer of bacteria.[11] Experiment 6-1 tested hand dryers installed in public locations, and Experiment 6-2 tested how far commercial hand dryers can spread bacteria. Overall, these experiments demonstrated that microorganisms contaminate both the air inlet and operating controls of public hand dryers and that commercial hand dryers aerosolize and spread microorganisms into the area immediately surrounding the dryer.

EXPERIMENT 6-1: Bacteria on Public Hand Dryers

Materials and Methods

We sampled hand dryers in these public places: on the Clemson University campus (25 dryers), at gas stations (14), and in grocery stores (21), for a total of 60 electric hand dryers. We took air samples by placing an open petri dish filled with plate count agar 6 inches from the outlet nozzle of each dryer. We also took swabs of dryer surfaces, such as the activation push button and inlet air vents. Air velocity, air temperature, and relative humidity were recorded for each dryer using a handheld hygro-thermometer. Placement of the air sample petri dishes at a distance of 6 inches was based on information from a hand dryer manufacturer that said most people place their hands approximately 6 inches away from the dryer air exhaust.[12] The dryer was activated and the nutrient agar was left under the air exhaust for 30 seconds (one cycle). A second plate was exposed for another cycle immediately following the first.

Science Stuff Ahead

Immediately after sampling, the open air plates (Difco Plate Count Agar; Becton Dickinson) were covered, placed in clean plastic bags, and transported to the laboratory. There, they were incubated at 37°C for 48 hours before total bacteria colony-forming units (CFUs) were counted and recorded. The dryer activation button and inlet vent of each hand dryer were also swabbed for microorganisms with a 5.5 cm × 5.5 cm sterile gauze previously wetted with 20 mL of sterile 0.1 percent peptone water (0.1 percent weight to volume; Becton Dickinson) stored in a Whirl-Pak bag (Weatherby/Nasco Inc.). The gauze was squeezed to remove excess peptone water and used to swab either the intake vent or activation button using sterile latex gloves.

Each dryer was swabbed with a new piece of gauze using a single-pass swipe before being returned to the bag and transported to the lab. Gauze samples were mixed at 250 revolutions per minute (rpm) for 30 seconds (Seward Stomacher 400 circulator, Seward Inc.). Then the peptone water was aseptically expressed from the gauze, removed from the bag, serially diluted with 0.1 percent peptone water, and plated in duplicate on Difco Plate Count Agar (Becton Dickinson). After incubation at 37°C for 48 hours, colonies were counted on dilution plates with 25 to 250 colonies on a Quebec colony counter and then converted to CFUs per sample. Statistical differences in mean bacterial populations were computed using SAS at a significance level of 5 percent.[13] The factors tested in Experiment 6-1 were public bathroom location, bathroom gender, and dryer air velocity.

Results of Experiment 6-1

In this experiment we found that the public hand dryers we sampled produced an average air velocity of 42 miles per hour (mph). Velocities of 15–89 mph were measured at air temperatures ranging from 190°F to 266°F and averaging 226°F. The open-air plate-sampling method detected bacteria coming out of 100 percent of the dryers. Total bacteria ranged from 2 to 238 CFUs/sampling cycle with a mean of 58 CFUs/cycle. No statistical differences in microbial populations on open-air plate samples were identified between dryers sampled in restrooms on a college campus, in grocery stores, or in gas stations. Bacteria recovered on open-air plates from gas stations averaged 64 CFUs/cycle, while those collected from grocery stores and college campus restroom dryers averaged 83 CFUs/cycle and 53 CFUs/cycle, respectively (Figure 2).

Activation push buttons and air inlets were found to be even dirtier. Gauze swab samples from air dryer inlets in grocery stores had significantly higher levels of microorganisms (18,620 bacteria/swab) than those from dryers located on a college campus (4,040 bacteria/swab) or dryers in gas stations (2,160 bacteria/swab). The reason for higher bac-

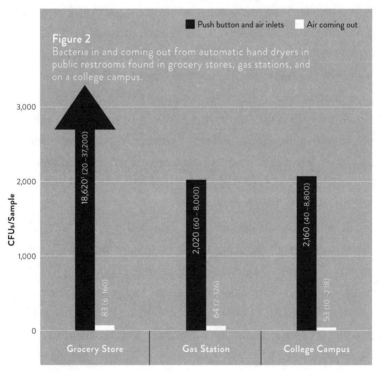

Figure 2
Bacteria in and coming out from automatic hand dryers in public restrooms found in grocery stores, gas stations, and on a college campus.

[1] Mean followed by the range of bacteria found in the samples. (n = 50 for college campus, 28 for gas stations, and 42 for grocery stores). Average air velocity and temperature were 18.7 m/sec (ranging from 6.8 – 40 m/sec) and 108.2°C (ranging from 88 – 130°C), respectively.

terial counts on hand dryers found in grocery stores is difficult to explain, but it may be associated with several factors such as more usage, dryer age, quality of dryer cleaning/maintenance, and so on.

Men's Bathroom Dryers Are Dirtier than Women's   Of all the dryers we sampled, 29 dryers (48 percent) were from men's restrooms and 31 dryers (52 percent) were from women's restrooms. The average number of bacteria detected on open-air sample plates from dryers located in men's restrooms was 125 CFUs/cycle, compared to 81 CFUs/cycle for dryers located in women's restrooms (Figure 3). These measurements were not, however, statistically different. Phew.

125 CFU/CYCLE   81 CFU/CYCLE

Okay ladies, you got us on this one. Bacterial numbers recovered from the activation push buttons and air inlets of dryers in men's restrooms were significantly higher as compared to the push buttons and air inlets on dryers in women's restrooms. We calculated means of 18,660 CFUs/mL for men's restrooms and 4,420 CFUs/mL for women's (see Figure 3).

When all data on inlets and outlet vents were combined, the microbial populations ranged from 20 to 37,200 CFUs/mL with a mean of 11,500 CFUs/surface. There were also more bacteria on air inlet vents than on the activation push buttons, likely because the outlet vents are pulling in air from the bathroom (mean for vents = 8,450 CFUs/mL; mean for buttons = 3,050 CFUs/mL), which contains airborne bacteria.

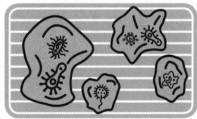

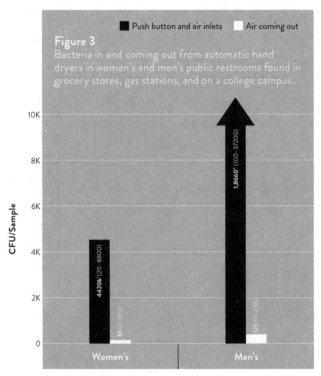

Push button and air inlets | **Air coming out**

Figure 3
Bacteria in and coming out from automatic hand dryers in women's and men's public restrooms found in grocery stores, gas stations, and on a college campus.

CFU/Sample

10K
8K
6K
4K
2K
0

1,8660* (100-37200)

4420b (20-8800)

81 (2-160)

125 (11-238)

Women's | Men's

a,b Means with different letters are significantly different at a level of 5%.
¹ Range of bacteria found in the samples shown in parentheses.
(n = 62 for women's bathrooms and 58 for men's bathrooms). Average air velocity and temperature was 18.7 m/sec (ranging from 6.8 – 40 m/sec) and 108.2°C (ranging from 88 – 130°C), respectively.

In addition, if these hand dryers are not cleaned regularly, air particulate matter and bacteria can be expected to build up on the air inlet surfaces. Have you ever noticed how much dust buildup you have on the front edges of those overhead room fan blades, the air inlets along the bottom of your refrigerator, or the surfaces of your furnace filters? Basically, the same principle applies: the more air that passes over these surfaces, the more debris and bacteria they capture.

Materials and Methods

Drs. Julie Northcutt and Michelle Parisi led this part of our blower study testing two commercial hand dryers: World Dryer (World Model A5-974 Dryer, World Dryer) and Fastdry Hand Dryers (Fastdry HK1800PS Hand Dryer, Allied Hand Dryers and Baby Changing Stations). Open petri dishes containing plate count agar were placed in various locations near the dryer to catch any airborne bacteria.

Science Stuff Ahead

Dryers were first placed in a sterile microbiological cabinet hood. Before sampling, the dryers were cleaned thoroughly around the air outlet nozzle and air intake vent using gauze pads saturated with 70 percent ethanol, which was allowed to air-dry for 5 minutes. The outlet nozzle was then inoculated with 0.1 mL of a suspension containing 10^7 to 10^8 cells per mL of our fluorescently labeled *E. coli* test strain (non-pathogenic *E. coli* strain JM109 was labeled with a jellyfish green fluorescent protein). This was the same *E. coli* used for the studies described in Chapters 3 and 5. The culture was spread across the nozzle tip using a sterile sample loop and allowed to air-dry for 20 minutes. Open petri dishes contain-

ing plate count agar were placed in nine different positions around the hand dryer. The dryer was activated for one 30-second cycle, after which the petri dishes were immediately covered.

The experiment was repeated three times, and a duplicate plate was tested each time following the first 30-second cycle. All plates were incubated at 37°C for 48 hours before CFUs were determined by counting fluorescent colonies under UV light and then multiplying those counts by the appropriate dilution factor. The data were analyzed to determine if there were differences in the total CFUs among dryer types and distances from the dryer nozzle at the 5 percent significance level using SAS.[14] For Experiment 6-2, we tested dryer type and locations from the dryer exhaust (12, 24, 48, or 60 inches from right under the dryer nozzle, and at angles of 0, +45, −45, +26, and −26 degrees coming straight out from the nozzle). Mean air velocity was 338 mph for the World Dryer and 280 mph for the Fastdry Dryer. Mean air temperature was 140°F for the World Dryer compared to 124°F for the Fastdry Dryer ($P = 0.01$).

Results of Experiment 6-2

After finding that hand dryers located in public bathrooms were in fact contaminated with bacteria, we also determined that they blow bacteria at least 1 yard away from the outlet nozzle. Among the nine plates placed in front of the dryer for 30-second cycles, the fewest bacteria found were on a plate 1 yard away, while the highest number were on a plate 1 foot away (Figure 4). However, bacteria were detected on 100 percent of the plate count agar samples at all distances, tested for both the World Dryer and the Fastdry model dryers. Statistically significant but relatively weak coefficients of determination were found between CFU recovery and dryer air velocity ($r^2 = 0.315$; $P = 0.0001$). Thus, as you might expect, the harder the dryer blows, the farther it blows bacteria.

Figure 4

Average number of bacteria (range in green) recovered from agar plates placed at different distances from a hand dryer inoculated with *E. coli* and then allowed to run one 30-second cycle

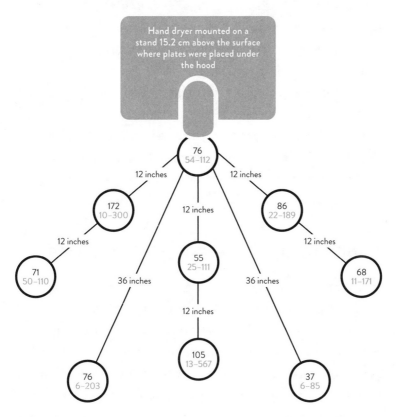

The higher air temperature of the dryer did not affect the number of bacteria being deposited around the room, at least in our study.

Other studies have shown that using electric hand dryers may increase numbers of bacteria on hands and on other surfaces.[15] Findings from Experiments 6-1 and 6-2 are in agreement with other research and demonstrate that electric hand dryers may promote bacterial transfer to other surfaces.[16] There was 100 percent incidence of bacterial recovery from both the open petri dish and sterile gauze swab sampling methods in public restrooms in Experiment 6-1. This finding indicates that dryer activation push buttons and air inlet vents of dryers in public restrooms are highly contaminated, even without a visible source of nutrients for bacterial survival and growth. Moreover, findings from Experiment 6-2 demonstrate that bacteria are transferred through the air from electric hand dryers and that the population of bacteria spread during each dryer cycle was similar (see Figure 4). All of this leads us to say that when using electric hand dryers, many people may be receiving bacteria that were left on or in the dryer by previous patrons of these so-called bacteria blowers. Gross.

THINGS TO CONSIDER: COLLATERAL DAMAGE!

To see how electric hand dryers spread water droplets, researchers from the Microbiology Department at the University of Leeds covered test participants' hands with black water-based paint and then dried their hands using both a jet-air and warm-air dryer with lower velocity. This and another follow-up study found that water droplets were spread over most of the user's body (Figure 5) and that yeast cells were spread up to 1.5 yards from a hand dryer after someone whose hands were inoculated with yeast used the dryer. In these studies, the jet-air dryer forced air out of the nozzle at 373 mph while the warm-air dryer had an air velocity of only 45 mph, yet both dryers also spread droplets and yeast cells to bystanders near the dryer. The microorganisms could be coming from

Figure 5
Number of paint spots on a white coverall worn by a participant whose hands were first dipped in black paint and then held under a jet-air dryer (adapted from Best et al., 2014)

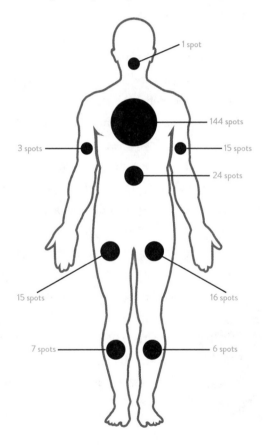

1 spot

144 spots

3 spots

15 spots

24 spots

15 spots

16 spots

7 spots

6 spots

the air outlet nozzle, air intake vent, and other unit components, or they could be simply drawn in from the bathroom air and redistributed into the room by the dryer. In either case, these units do transfer bacteria.

Electric hand dryers are not recommended for healthcare settings, and this raises the question of why they are used in food retail and food processing establishments. Rather than minimizing cross-contamination, hand dryers are likely to transfer bacteria to clean hands, worker clothes, or surrounding surfaces. Hand dryers may increase the spread of foodborne illness and disease-causing microorganisms to individuals and foods prepared and consumed in these environments, and even higher levels of contamination could occur if workers are ill or carrying infection. The World Union of Wholesale Markets at the European Tissue Symposium recommended that because air dryers and roller towels may induce microbiological contamination, they should not be used in food preparation rooms.[17] Thanks to a 1-yard minimum spread of aerosolized bacteria, as found in Experiment 6-2, workers and surfaces are at risk for contamination even if they are not directly using these dryers.

The Leeds University study highlights the importance of appropriate drying methods in food processing and retail establishments. In addition to the concern over the potential spread of microorganisms via air hand dryers, observational studies have found that the average drying times for men (17 seconds) and women (13.3 seconds) do not meet the designated 30 seconds for the dryer cycle.[18] Furthermore, hand wetness was found to play a significant role in transfer of bacteria for food[19] and hospital settings.[20] Hygiene education is critical, and it calls for

You want aerosolized bacteria with your burger?

new strategies that motivate workers to follow proper techniques for hand washing and drying. While cleaning and sanitizing electric hand dryers may reduce the risk of aerosolizing potentially harmful microorganisms, we did not test the effectiveness of such cleaning methods in our study.[21] If hand dryers are not suitable for the health-care industry due to the chance of spreading germs, this might indicate that air hand dryers are not appropriate for food-handling environments.

TOWELS VERSUS AIR DRYERS

Spoiler Alert: Towels almost always win. Based on the Mayo Clinic review, let's look at the scorecard.

 Drying efficiency. Paper towels and jet-air dryers achieved 90 percent dryness in about 10 seconds, while warm-air dryers took 40 seconds to achieve the same level of dryness.

 Removing bacteria. Jet-air dryers were better than warm-air dryers (especially with UV light), but paper towels were slightly better than jet-air dryers, especially in removing bacteria from fingertips.

 Cross-contamination. Before we get back to hand-drying methods, let's consider the airborne bacteria within the bathroom zone. Each time a urinal or toilet is flushed, a mist is sprayed into the air surrounding the unit for up to 6 yards in each direction.[22] You probably don't want to think too much about this potential "fecal/urine mist" when you visit a public lavatory, but there is some concern that air hand dryers further contribute to the spread of bacteria. Warm-air dryers were found to spread bacteria up to 3 feet, while jet-air dryers spread bacteria up to 6 feet from the dryers.[23] Paper towels, however, were found not to spread bacteria.

 Preference. In three separate surveys in the UK, United States, and Australia, people preferred to use paper towels over air dryers by 55 to 60 percent.[24]

 Noise. Obviously, air dryers generate more noise than paper towels do. Jet-air dryers generated 92–94 decibels within 2 yards of the unit, which is similar to a large truck passing within 3 yards.

 Cost. Paper towels are more costly, since they need to be replaced and require waste disposal. On the other hand, the initial investment for air dryers is higher, ranging from ~$100 to $1,800. Paper towels cost from 0.75¢ to 2.2¢ per sheet, and towel dispensers are from $15 to $50 each.

 Environmental effect. Here the air dryers come out on top. A life-cycle assessment is a technique for determining the environmental impact associated with all steps in a product's life, from raw material to disposal. Researchers from the University of Guelph concluded that using two paper towels to dry your hands has a greater environmental impact than using a warm-air or jet-air dryer for the same task.[25]

Although other research from the Mayo Clinic found that using electric hand dryers and paper towels reduces about the same number of bacteria,[26] our evidence leans toward electric air dryers for reduced waste and energy usage, and most of the other research supports paper towels as the better bacteria reducer. Besides, even if you use an electric hand dryer, you probably wipe your hands on your pants anyway. Which brings us to our next set of studies—transport mechanisms.

part
3

Transport

We borrowed the title of this part from Chef Alton Brown, host of *Good Eats* on the Food Network—it's his description of a dipping chip. In that episode, Chef Brown defined a dip based on its ability to "maintain contact with its transport mechanism over three feet of white carpet."[1] In Part 3 we'll look at different ways to transport bacteria onto and into the food we eat: from handling food with hands, utensils, and chopsticks to handling lemon slices and popcorn, and finally, to double-dipping a chip.

The main issue here is how microorganisms are transferred between people. As many people know, transfer of microorganisms occurs in various ways—touching, body fluids (including saliva), air, food, water, insects, and fomites (nonliving objects such as bedding, toys, and almost any inanimate surface). Most of us also know that microbes transfer by air (sneezing, coughing, electric hand dryers), water (drinking water, contaminated irrigation, or rinse water), and food. Insects also transfer pathogens; for example, mosquitoes transmit malaria, rat fleas spread the bubonic plague, and flies carry bacteria to your food on their legs. Fomites include towels or floors that spread athlete's foot fungus and cutting boards harboring salmonella that gets on your washed lettuce. Microorganisms can also move from one person to the next by catching a ride on a lemon slice held for cutting, a spoon after it's been used, or a popcorn kernel touched by a hand that's just been in someone's mouth. And of course we have to include a chip that has been double-dipped. So we could probably call these items fomites as well.

We humans are constantly sharing microorganisms. Usually there are no harmful effects, and sometimes we benefit from the results, which can include building immunity. Exposure to bacteria at an early age helps strengthen our immune system by developing a diverse and healthy mixture of bacteria in the gut. Also, humans have an adaptive immune system that gives us "immunological memory" and thus can protect us from pathogens after being exposed to the pathogen or after a vaccine that mimics the pathogen. Once exposed to a pathogen, the body produces specific defense agents (antibodies, also called immunoglobulins) that attack the pathogen.

Vaccines have been developed that use this natural mechanism to build immunity to chicken pox (varicella), polio (poliovirus), rubella virus (also known as German measles; not the same as measles), measles (rubeola virus), mumps virus, and smallpox (variola viruses, of which there are two). Some of these viruses can have serious consequences, including death. Vaccines have been credited with eradicating smallpox, since the last known naturally occurring case was in 1977. A vaccine has also been credited with greatly reducing the incidence of typhoid fever, which is caused by a bacterium (*Salmonella* Typhi). So, it is true that exposure to pathogens can help develop immunity to disease. Perhaps taking it a bit too far, some parents are deliberately exposing children to chicken pox at "pox parties," which the medical field deems unsafe.[2]

Still, some in the medical field have come up with the so-called hygiene hypothesis, which states that exposure to parasites, bacteria, and viruses at a young age strengthens the immune system and reduces the chance of having allergies, asthma, and other inflammatory problems.[3] The hygiene hypothesis states that immunity to specific disease agents develops by exposure to those pathogens, since the body manufactures antibodies specific to these agents. However, others in the medical field believe that an overall strong immune system is developed by exposure to specific "good" bacteria.[4]

The immune system can be overwhelmed by disease agents, of course, and people still get sick from the common cold, flu, and other pathogens sometimes associated with food. So while it may be important to expose yourself to microorganisms and thus strengthen your immune system, it's not always a good idea. As mentioned earlier, immunocompromised people are at particular risk of infection by pathogens, and even the healthiest of us can fall into this category if we are experiencing physical or mental stress or are already fighting off a cold.[5]

But let's get back to transport mechanisms: microorganisms on people's hands and in their mouths can, in fact, catch a ride on various items and travel from person to person. Most of these oral bacterial communities reside on the teeth and eventually develop into plaque or biofilms. Human oral bacteria are not naturally found anywhere else in the body and may include microorganisms that can cause contagious diseases.[6] Oral bacteria can include pathogenic strains of bacteria and viruses, and disease is spread via mucus.[7] In fact, complex statistical models have been developed to predict the spread of disease from a single individual to large populations.[8]

The overall impact of disease transmission from food and food "fomites" touched by others is probably underestimated. It's difficult to know if the cold you picked up came from sharing a bag of popcorn. Each year 1.5 million people die from respiratory infections and 1.7 million die from diarrheal infections around the world, and 33 percent of respiratory infections and 94 percent of diarrheal infections are due to environmental factors.[9] Transmission by fomites has been proven for ten of the most common viral diseases and pathogenic bacterial strains.[10] We already know hands are a major way we transfer disease in health-care settings and foodborne transmissions.[11]

Pathogenic microorganisms can contaminate food at any point from harvest to consumption. They can be transferred to food prepared by an infected person (person-to-person spread), transmitted through the air, and carried by insects, pests, rodents, or pets. In some cases, the disease-causing microorganisms can remain with an infected person even after recovery. A person with this condition is known as a carrier. One of the most infamous carriers was a cook named Mary Mallon, nicknamed "Typhoid Mary." In the early twentieth century in New York, she was identified as a chronic carrier and transmitter of typhoid fever. Typhoid Mary displayed no overt symptoms of disease, but she was notorious for spreading typhoid fever (*Salmonella* Typhi). Using genome sequencing, this bacterium has been more precisely named *Salmonella enterica* subspecies *enterica* serovar Typhi.

In the next few chapters, we look into how microbes move from person to person in less expected ways, such as when someone puts lemon slices in your drink, friends are sharing a bowl of soup, and partygoers do the infamous double-dip.

7

THINGS YOU PUT IN YOUR DRINK

Chapter 7

THINGS YOU PUT IN YOUR DRINK

Because everyone consumes drinks, both alcoholic and nonalcoholic, they are a large component of the food industry. For 2015 there was an $841 billion global soft drink market and a $1,344 billion alcoholic beverage market.[1] The Party Drink Calculator states that, on average, each guest at a party will consume two beverages the first hour of the party and then one drink every hour thereafter.[2] In his day, William Claude Dukenfield (stage name W. C. Fields), a well-known actor in the twentieth century, did not set a good example of responsible drinking. Famous for his por-

I drink, therefore I am

trayals of an inebriated character, he was born too soon to follow the Party Calculator for his consumption rate.

What do you add to your drink? Drinks such as iced tea are often prepared with a slice of lemon or by adding a handful of ice. Sometimes a server handles the fruit and ice, sometimes it's the drinkers themselves; but people often use their hands or an ice scoop, both of which offer ample opportunity for contamination. We've now determined, of course, that dirty hands are a major player in transmitting disease via food.[3]

 ## ON THE ROCKS, PLEASE

What about ice? You might not think about it much, but those little cubes of ice that you put in your cocktail or soda can be contaminated with the pathogenic organisms we talked about earlier and can thus contribute to spreading illness.[4] In 1987, for example, an ice-borne disease caused an outbreak of Norwalk virus (norovirus) in Pennsylvania, Delaware, and New Jersey. The Centers for Disease Control and Prevention (CDC) estimated that more than five thousand people were affected.[5] In Peru, contaminated ice was also responsible for the 1991 cholera epidemic, which caused 7,922 illnesses and 17 deaths in Peru and then spread throughout Latin America.[6] In the early 2000s, diarrheagenic *E. coli* were detected in commercial ice produced in Brazil.[7] In 2005, ice produced at retail outlets in Nigeria was contaminated with more than 1,000 antibiotic-resistant bacteria per milliliter (isolates displayed 100, 67, and 87 percent resistance to ampicillin, erythromycin, and tetracycline, respectively).[8] In other examples, pathogens have been detected in ice from ice-making machines.[9]

In a survey of over 3,500 samples of ice used to cool drinks, 9 percent contained coliforms and 11 percent had total bacterial counts of greater than 1,000 bacteria/mL.[10] For those who like their upscale coffee, in 2017, Starbucks, Costa, and Caffé Nero iced coffee drinks were found to contain fecal bacteria.[11] Sorry about that!

LEMONS

That bright, cheerful-looking citrus fruit we call a lemon can be quite helpful when it comes to fighting off bacteria. Various studies have found that lemons and lemon extracts will kill many types of bacteria, yeasts, and molds, including foodborne pathogens such as *E. coli, Listeria*, and *Salmonella* species.[12] Lemon juice has even been found to inhibit HIV.[13] But lemons also naturally carry bacteria from the environment, meaning that cross-contamination can occur between lemons, hands, and the other thing we put into our drinks—ice.[14] What's more, we know that bacterial transfer to foods from hands is particularly high when you touch your fingers to your lips.[15] (Maybe you're doing that right now while reading this book. If so, we'd suggest curbing that habit.)

When bacteria are on a fruit's skin, or peel, they can end up in whatever beverages you might make from that fruit. For example, orange peel inoculated with *Salmonella* Typhimurium, *E. coli*, and *Listeria monocytogenes* ended up in the juice made from those same oranges.[16] Bacteria might also get into your drink from a lemon slice floating in your beverage, or from one placed on the rim of your glass as garnish. We've all seen those little cut-up lemon wedges at self-service drink stations. Well, the restaurant workers aren't the only ones who touched those lemons—other customers may have been plunging their unwashed hands into the bowl and grabbing onto a slippery citrus garnish. With what we know about bacteria transfer so far, we can say that lemon slices are exposed to many opportunities for contamination, and the ones that sit unrefrigerated at drink stations might just support enough microbial growth to make you sick after picking one up.

PEOPLE HANDLING YOUR FOOD

Cross-contamination in food service plays an important role in foodborne illness.[17] During food prep, bacteria on hands can be transferred directly to raw foods from hands and indirectly from other surfaces the food touches.[18] Food handlers who may be ill and shedding pathogens can also be a source of contamination even though they are showing no overt symptoms.[19] (Remember the case of Typhoid Mary?) At least thirteen different studies have examined the transfer of bacteria to food from food contact surfaces including stainless steel, fabrics, gloves, and hands.[20]

To determine the truth for ourselves, we identified three objectives for our experiments. We wanted to find answers to these questions:[21]

1. To what extent are bacteria transferred to lemons when handled with contaminated hands?
2. Do bacterial populations increase during the storage of contaminated lemons?
3. How extensive is bacterial transfer to ice when handled with contaminated hands or scoops?

EXPERIMENT 7-1: *E. coli* Transfer from Hands to Wet Lemons versus Dry Lemons

Materials and Methods

As in previous experiments, we used an ampicillin-resistant *E. coli* strain containing a fluorescent gene to track the transfer of bacteria from hands to lemons. All partici-

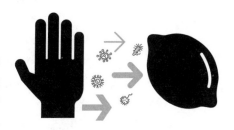

pants washed their hands with warm water and soap and allowed their hands to air-dry. Then we deposited 1 mL of the *E. coli* inoculum of about 1 million cells per mL into the center of the palm of their dominant hand. Participants applied the *E. coli* by rubbing their hands together for 30 seconds, and then allowed their hands to air-dry for 30 seconds. Participants then handled a lemon for 30 seconds by rolling the lemon between their hands. We tested both wet lemons and dry lemons, comparing the number of *E. coli* transferred to each lemon with bacteria recovered when participants handled the lemons with non-inoculated hands.

Science Stuff Ahead

After participants handled the lemons, both hands and lemons were massaged for 30 seconds in separate bags containing 20 mL of sterile 0.1 percent peptone water to facilitate recovery of bacteria remaining on the hands and the lemons. Next, 1 mL of the peptone water cell suspension was removed from each stomacher bag, placed into 9 mL of sterile 0.1 percent peptone water, and then serially diluted. An aliquot of 0.1 mL from the sample dilutions was pipetted and spread onto the surface of duplicate tryptic soy agar plates. Plates were held for 5 to 10 minutes and then inverted and placed in an incubator at 37°C for 24 hours. The following day, the plates were inspected under UV light. Plates with 25 to 250 CFUs/plate of fluorescent-appearing colonies were counted, multiplied by the dilution factor, and then that number was converted to CFUs/hand and log CFUs/hand.

The percent transfer of *E. coli* from hands to lemons was calculated using this formula:

$$\% \text{ transfer} = \frac{\text{CFUs recovered from lemons}}{\text{CFUs recovered from hands} + \text{CFUs recovered from lemons}} \times 100$$

This method of calculating percent transfer overestimates the transfer amount, since all of the bacteria that hands were originally inoculated with will probably not be recovered. At any rate, not all of the bacteria originally inoculated may be available for transfer (that is, physically attached to the hands or lemons). For the hand-to-lemon transfer, the difference in transfer was about 1 percent more for the method shown in the formula above than for the method using the total inoculum population as the denominator.

Descriptive statistics were determined for this experiment, as well as the next two, using the Statistical Analysis System (SAS).[22] In Experiment 7-1 (bacterial transfer from hands to lemons), there were 11 to 13 test participants for all three replications, and each observation was duplicated for a total of 70 observations per treatment (using wet or dry lemons). Experiment 7-2 (bacterial transfer from hands or scoops to ice) employed 11 participants for all three replications, and each observation was duplicated for a total of 66 observations per treatment (hand or scoop). Experiment 7-3 (bacterial survival on stored lemons) examined two treatment variables: (1) the storage temperature (room or refrigerated) and (2) the storage time (0, 4, and 24 hours). Duplicate samples were used for each of five replications, for a total of ten observations for each treatment of combined storage temperature and storage time. Treatment effects were tested to determine differences in *E. coli* counts at a 5 percent significance level using SAS.[23]

Results of Experiment 7-1

E. coli was transferred to 100 percent of the lemons that were wet before handling with inoculated hands, while only 30 percent of the dry lemons had detectable *E. coli*. The average number of bacteria per wet lemon was 6,123; an average of 5 percent bacteria transferred from the hands (Figure 1). Conversely, the dry lemons picked up an average of 1,502 bacteria, which accounted for bacterial transfer of 0.7 percent for the 30 percent of the lemons contaminated with *E. coli*.

Common sense tells us—and previous research has concluded—that wet surfaces pick up more bacteria than dry ones.[24] For example, one such study reported greater transfer of *E. coli* from meat to gloves and from gloves to meat when the gloves were wet.[25]

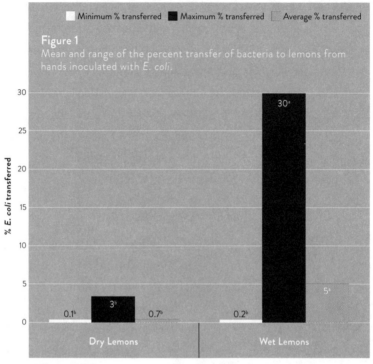

Figure 1
Mean and range of the percent transfer of bacteria to lemons from hands inoculated with *E. coli*.

% *E. coli* transferred

30
25
20
15
10
5
0

Dry Lemons
0.1[b]   3[b]   0.7[b]

Wet Lemons
0.2[b]   30[a]   5[a]

a,b Means with different letters are significantly different at a level of 5%; $n = 70$.

EXPERIMENT 7-2: *E. coli* Hanging Out on Sliced Lemons

Materials and Methods

Survival of *E. coli* on lemon slices was tested at three different times (0, 4, and 24 hours) and at refrigerated (4°C ± 2°C) and room (21°C ± 2°C) temperatures. Lemons were inoculated with the fluorescent ampicillin-resistant *E. coli* by placing each lemon in a sterile bag containing 20 mL of the *E. coli* solution. The bag was shaken for 30 seconds, after which the lemon was removed and allowed to air-dry for 5 minutes. The inoculated lemons were then sliced into quarters. *E. coli* populations were enumerated for one set of lemon quarters after 10 minutes; the other lemon quarters were stored for 4 and 24 hours at either room or refrigerated temperature. *E. coli* populations were recovered from the lemon quarters and enumerated after each storage time using the method described in Experiment 7-1.

Results of Experiment 7-2

As you might expect, the *E. coli* was most prominent right after inoculation. However, the number of bacteria on lemons was higher after 24

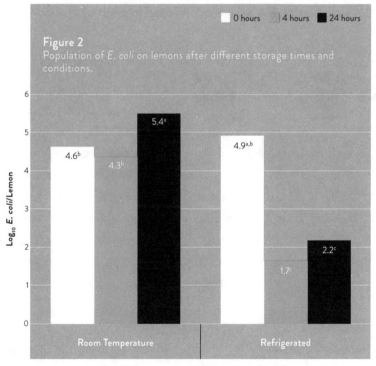

☐ 0 hours ▨ 4 hours ■ 24 hours

Figure 2
Population of *E. coli* on lemons after different storage times and conditions.

Room Temperature: 4.6[b] (0 hours), 4.3[b] (4 hours), 5.4[a] (24 hours)
Refrigerated: 4.9[a,b] (0 hours), 1.7[c] (4 hours), 2.2[c] (24 hours)

Y-axis: Log_{10} *E. coli*/Lemon

a,b,c Means with different letters are significantly different at a level of 5%; n = 10.

hours when held at room temperature (Figure 2). Refrigeration reduced *E. coli* populations from about 100,000 to about 100 CFUs/lemon after 4 hours, but that number did not further decrease after 24 hours of refrigerated storage. E. Fernández Escartin and his coworkers from the University of Guadalajara found that populations of pathogens like *Salmonella* and *Shigella* increased on cut watermelon and papaya within 6 hours at room temperature, and even after lemon juice was applied to the surfaces of cut fruit, the bacterial populations increased after 2 hours.[26]

EXPERIMENT 7-3: *E. coli* Transferred to Ice from Hands and Metal Scoops

Materials and Methods

Participants washed their hands with warm water and soap and allowed their hands to air-dry. Then we deposited 1 mL of the fluorescent and ampicillin-resistant *E. coli* inoculum into the center of each participant's dominant hand. As in Experiment 7-2, *E. coli* was distributed to the participants, who then rubbed their hands together for 30 seconds and air-dried their hands for 30 seconds. The sanitized metal scoops were inoculated by placing 1 mL of the *E. coli* in the center of the scoop, spreading it across the scoop surface using a sterile glass rod, and allowing the scoop to air-dry for 30 seconds.

Science Stuff Ahead

We tested four treatments to determine the extent of bacterial transfer from hands or scoops to ice:

1. Non-inoculated hands handling ice
2. Inoculated hands handling ice
3. Non-inoculated scoop carrying ice
4. Inoculated scoop carrying ice

All participants washed and air-dried their hands. Then they picked up a handful of ice and immediately placed the ice in a filter stomacher bag containing 20 mL of sterile 0.1 percent peptone water. This procedure was repeated with inoculated hands and with both inoculated and non-inoculated scoops. The ice and peptone water were mixed for 30 seconds in the bag. To enumerate bacteria on a participant's hand, the dominant hand used to pick up the ice was placed into a sterile stomacher bag containing 20 mL of sterile 0.1 percent peptone water and agitated for 30 seconds, covering all fingers, the palm, and the back of hand. Next, 1 mL of the peptone water/cell suspension was removed from each stomacher bag, placed into 9 mL of sterile 0.1 percent peptone water, and serially diluted in 0.1 percent peptone water. We then spread 0.1 mL of the sample dilutions onto the surfaces of tryptic soy agar plates containing 100 mg ampicillin/mL. Plates were held for 5 to 10 minutes, inverted, and then placed in an incubator at 37°C for 24 hours. On the following day, the plates were inspected under UV light, and plates with fluorescent colonies ranging from 25 to 250 CFUs/plate were counted and converted to CFUs/mL based on the appropriate dilution. Bacterial colony counts were converted to CFUs/hand or scoop and log CFUs/hand or scoop based on the amount of rinse solution used. Percentage of *E. coli* transferred from hands to ice was calculated as follows:

$$\% \text{ transfer} = \frac{\text{CFUs recovered from ice}}{\text{CFUs recovered from hands (scoops)} + \text{ice}} \times 100$$

Results of Experiment 7-3

In our study, an average of 19.5 percent of the bacteria on the hands and 66.2 percent of the bacteria on the scoops were transferred to ice (Figure 3). When we based the percentage of bacteria recovered on the total number of bacteria inoculated onto hands rather than the population of bacteria we could recover after inoculation, the percent transfer was about 2 percent for hands and 40 percent for scoops. This number was just under 20,000 bacteria transferred from the hands and around 500,000 bacteria transferred from the scoops to the ice using either method to calculate percent transfer. The higher level of transfer from scoops compared to hands is expected because the smoother and less porous surface of stainless steel is less conducive to bacterial attachment than skin is.

Previous research has found that bacteria residing on surfaces could shed during repeated contact with other surfaces.[27] Ice, in particular, has a long history of transmitting pathogenic microorganisms to humans.[28] In many of these cases, the pathogen was carried in water used to create the ice; however, cross-contamination due to physical handling of food and ice has also caused sickness.[29] Numerous studies have reported the transfer of bacteria to food from hands as well as from stainless steel surfaces.[30]

Bacteria also can transfer from contaminated ice-holding bins or scoops to ice. One study found *E. coli* in 6.7 percent of ice samples and 22 percent of ice chest samples, even though the water used to make the ice contained no *E. coli*.[31] Other studies have found a variety of pathogens in ice, including *Salmonella*, fifty different *E. coli* strains, and bacterial spores.[32]

Researchers in yet another study found bacterial transfer rates

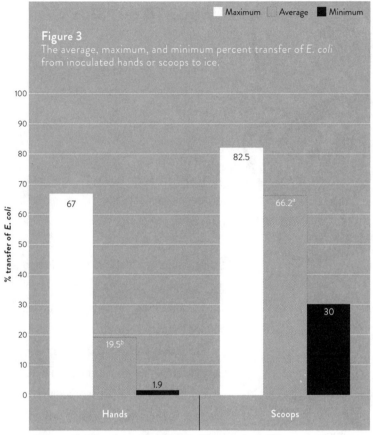

Figure 3
The average, maximum, and minimum percent transfer of *E. coli* from inoculated hands or scoops to ice.

% transfer of *E. coli*

100
90
80
70
60
50
40
30
20
10
0

67
19.5[b]
1.9

82.5
66.2[a]
30

Hands

Scoops

a,b Means with different letters are significantly different at a level of 5%.

to lettuce as high as 100 percent even when hand washing occurred before handling the lettuce.[33] To further complicate the issue, this study reported that touching other surfaces such as spigots creates "hot" surfaces that become sources of contamination. As we know and have seen, cross-contamination in food-service environments is a major factor in many outbreaks of foodborne illness.[34]

THINGS TO CONSIDER

Could I have a little privacy, please?

Refreshing cocktail! You guys gotta try this!

"I never drink water because of the disgusting things fish do in it."

—W. C. Fields

A common fallacy is that alcoholic and acidic beverages, such as lemon juice, will kill bacteria and protect you from harmful microorganisms. In fact, one study found that adding lemon to water actually increased the bacterial growth.[35] As for alcohol, it did not kill off pathogens such as *Shigella sonnei* and *Shigella flexneri, Salmonella* Typhi, and *E. coli* that were frozen in ice and allowed to melt for 30 minutes in cola, soda, Scotch whisky (80 proof), a Scotch/soda mixture, and tequila (86 proof).[36] The following bacterial recovery percentages were based on the number of bacteria inoculated into the ice compared to those recovered from these beverages after the ice melted in each type of drink: 100 percent in club soda, 55–74 percent in cola, 64–94 percent in Scotch and soda, 11–16 percent in pure Scotch, and 5–10 percent in pure tequila. In a study to determine the best beverage to consume to avoid so-called traveler's diarrhea or Montezuma's revenge, neither diet cola, beer, regular cola, nor sour mix completely eliminated *Salmonella* Typhimurium and enterotoxigenic *E. coli* after 24 hours of exposure; however, wine eliminated these bacteria within 4 hours.[37] (Go vino!)

I'm just here to help.

And how about lemons? One study found that 69.7 percent of lemons collected from drinks in twenty-one restaurants were carrying microorganisms, including many types of bacteria associated with human contamination.[38] When orange drinks with acidic pH levels of 3.0, 4.9, and 6.8 were examined, only the pH 3.0 drink reduced *E. coli* and *Salmonella* species (spp.) populations, but low levels were not observed until after 30 hours of exposure.[39] The extremely long exposure times tested are not likely to occur in drinks served with ice or freshly cut lemons. Thus we can say that the contamination of ice or lemons could lead to the transfer of pathogens to drinks and ultimately to you. Food-service workers are the primary source of contamination of food with norovirus, hepatitis A, and *Shigella* spp., and they can transfer other pathogens such as *E. coli* and *Salmonella* spp. to the food they prepare or bring to your table.[40] So next time you order a club soda with a wedge of lemon, think twice about the cleanliness of the bar *and* your bartender's hands.

HANDLE WITH CARE

8

CAN I HAVE A TASTE OF THAT??

Chapter 8

- - - - - - - - - - - - - - - - - - - -

CAN I HAVE A
TASTE OF THAT?

"**D**o you want a taste of my soup?" "Can I have a sip?" "Let's order several different items and share them." You've no doubt heard these phrases being uttered around a table of friends eating a meal. Sharing food with others is a social activity that gives a sense of belonging and even intimacy. The social significance of sharing food has been studied by psychologists in humans and animals alike. Food sharing by college students has social significance, and consubstantiation (sharing food that has been touched, tasted, or bitten) implies a positive and friendly social relationship and increases feelings of intimacy and closeness.[1] (Bro, are we consubstantiating?) Consubstantiation also has a theological meaning, but psychologists have adopted a broader interpretation that relates intimacy with sharing food. In fact, sharing a meal was found to decrease dominance and submissiveness while increasing agreeableness and bonding.[2] But does sharing food open the door for cross-contamination?

Sharing food is a common practice in many cultures. People eat directly from a communal bowl or plate using chopsticks, spoons, forks, and even hands placed in their mouths. In many cultures, people still eat by picking up food with their hands. For instance, eating rice and naan with hands is a common practice during Indian meals, and in Ethiopia, people use their hands to soak up sauces into *injera*, a spongy flatbread. The Spanish have tapas that are meant to be picked up and shared. In Vietnam, it is a sign of politeness to let your guests serve themselves to the communal dish before you serve yourself. Sometimes the host will even pick up food and put it into his guest's bowls. We've seen family-style serving become more common recently in the United States, both in restaurants and homes.

As we know, contaminated hands are a major source of contamination in any food-service area.[3] Thus, hands are also a source of cross-contamination that can then result in increased outbreaks related to foodborne pathogens, both bacterial and viral. People use their hands to share and eat food in many settings, such as when we consume snacks like popcorn (see Chapter 9), nuts, and candies at movie theaters, in pubs, and at home. In situations like these, people aren't washing their hands between mouthfuls, so they increase the likelihood of transferring microorganisms from person to person and potentially spreading disease. There's a long list of diseases that can be spread from the mucus or skin of an infected individual: the common cold, flu, meningitis, rubella, chicken pox, measles, tuberculosis (TB), cold sores, and staph infections as well as hand, foot, and mouth disease.[4]

So what are some of the ways people might exchange undesirable microbes while sharing a meal? Well, let's look at three scenarios:[5]

 Experiment 8-1: Sharing utensils (transfer of bacteria from the mouth to a utensil)

 Experiment 8-2: Sharing soup (transfer of bacteria from the mouth to broth)

 Experiment 8-3: Sharing rice (transfer of bacteria from the mouth to rice)

MATERIALS AND METHODS

We divided these scenarios into three separate experiments. Each experiment was replicated three times on different days. Utica fine quality 18/0 stainless steel teaspoons (Utica Fine Quality Cutlery Company) and disposable wooden chopsticks (Home Plus) were used as eating utensils. Utensils were sterilized before use. For food samples, we used Swanson's chicken broth (Campbell Soup Company) and Mahatma extra-long-grain enriched rice (Riviana Foods Inc.). Salt concentration of the broth was 4 mg/mL (based on the food label). To reduce the salt concentration, 20 mL of chicken broth was diluted with sterile water at a ratio of 1:4. The result was a salt concentration in the diluted broth of 1 mg/mL and a final volume of 80 mL used for testing. Mahatma extra-long-grain enriched rice was cooked according to the manufacturer's directions in a microwave oven (1,000 watt, Magic Chef, USA) for 15 minutes using a 1:2 ratio of rice to water.

STATISTICAL ANALYSIS

Each experiment was replicated three times. Experiments 8-1 and 8-2 had ten participants, and Experiment 8-3 included seven participants. All bacteriological plating was performed in duplicate, and all treatments were tested to determine differences at a 5 percent significance level using the Statistical Analysis System (SAS).[6]

EXPERIMENT 8-1: Sharing More than a Spoon

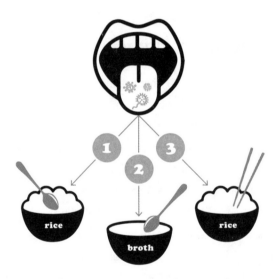

To determine if sharing utensils might pose a problem with cross-contamination, we compared the number of bacteria recovered from utensils when they were exposed to several different scenarios. The scenarios were recovery of bacteria from (1) utensil only, (2) utensil exposed to food (broth or rice), (3) utensil put in the mouth, and (4) utensil put in the mouth with food. The first two treatments are experimental controls, or treatments not exposed to what we are testing, which in this case is exposure to the mouth. Using appropriate experimental controls is important in this case to prove that the bacteria found on the utensil is from the mouth and not from another external source.

Science Stuff Ahead

Experiment 8-1 was divided into three parts to test the transfer of bacteria from the mouth to a utensil while (1) eating rice with a spoon, (2) eating chicken broth with a spoon, and (3) eating rice with chopsticks. For each experiment, ten participants (five male and five female) were trained to use a similar technique by placing the utensil containing the food on their tongue and completely into their mouth.

1. Eating rice with a spoon
   - **Just the spoon (spoon control).** A sterile teaspoon was placed in a sterile bag containing 20 mL of 0.1 percent peptone water (Difco Laboratories), then agitated and manually scrubbed for 30 seconds. A 1-mL aliquot was taken from the rinse solution, serially diluted, and pour-plated in duplicate using plate count agar (Difco Laboratories) and incubated at $37°C \pm 2°C$ for 48 hours. Plates from dilutions having 25 to 250 colonies were counted and then converted to log CFUs/20 mL of rinse.
   - **Just the rice (rice control).** This treatment was the same as the

spoon control except rice was placed on the spoon before recovering bacteria.

- **Spoon in the mouth (spoon/mouth).** A sterile spoon was aseptically inserted once into the participant's mouth without contact with rice. Bacteria recovered from the spoon were enumerated as described for the rice control.

- **Spoon in the mouth with rice (rice/spoon/mouth).** This treatment was the same as the spoon/mouth treatment except rice was placed on the spoon before putting the spoon in the mouth.

2. Eating chicken broth with a spoon consisted of the same four treatments, except heated chicken broth (as per manufacturer's directions) was used instead of cooked rice.

3. Eating rice with chopsticks also consisted of the same four treatments, which were identical to part 1 of this experiment except chopsticks were used instead of spoons.

Results of Experiment 8-1

We observed a transfer of over 100,000 bacterial cells to the spoon and chopstick when placed in the mouth, with or without food (Figure 1). Notice that we used a log scale on the graph for expressing the bacterial counts (bacterial populations are shown in non-log numbers). As mentioned in an earlier chapter, a log scale is used because it corresponds to the logarithmic growth of microorganisms and makes it easier to show differences in very large numbers. There was no statistical difference in bacterial transfer whether the utensil was put directly in the mouth or used to get food before being placed in the mouth

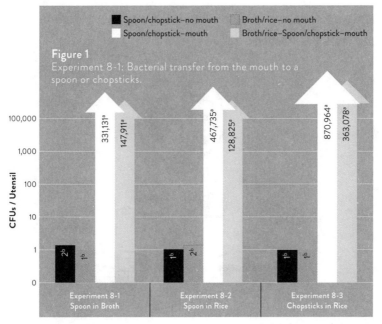

Figure 1
Experiment 8-1: Bacterial transfer from the mouth to a spoon or chopsticks.

Legend:
- Spoon/chopstick–no mouth
- Broth/rice–no mouth
- Spoon/chopstick–mouth
- Broth/rice–Spoon/chopstick–mouth

Y-axis: CFUs / Utensil (100,000 / 1,000 / 100 / 10 / 1 / 0)

Experiment 8-1 Spoon in Broth: 2[b], 1[b], 331,131[a], 147,911[a]
Experiment 8-2 Spoon in Rice: 1[b], 2[b], 467,735[a], 128,825[a]
Experiment 8-3 Chopsticks in Rice: 1[b], 1[b], 870,964[a], 363,078[a]

a,b Means within experiments with different letters are significantly different at a level of 5%; n = 10.
Spoon/chopstick–no mouth = bacteria recovered from the spoon or chopstick with no contact with mouth or broth.
Broth/rice–no mouth = bacteria recovered from the spoon with contact with broth or rice without placing in the mouth.
Spoon/chopstick–mouth = bacteria recovered from the spoon or chopstick after placing in the mouth without contacting the broth or rice.
Broth/rice–spoon/chopstick–mouth = bacteria recovered from the spoon or chopstick after first placing them in the broth or rice followed by the mouth.

(see Figure 1). The population of bacteria recovered from the controls (utensils after being placed into the food) was less than ten cells for both the spoon inserted into the broth or rice and for the chopsticks inserted into rice.

EXPERIMENT 8-2: It's Like Putting Your Mouth in the Soup

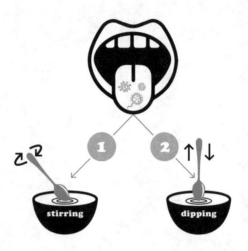

Materials and Methods

So, what about the second scenario of sharing a bowl of soup? We know that bacteria are transferred to the spoon after it contacts the mouth. What happens if you stick this same contaminated spoon back into the soup? In our experimental design, we examined if the way the spoon was inserted into the soup would have any effect on the population of bacteria transferred into the soup. Again, experimental control treatments were tested where the spoon was not placed in the mouth before being inserted into the soup. For example, controls with no mouth contact consisted of either just dipping the spoon in the soup or stirring the spoon in the soup. Thus the two treatments tested included inserting the spoon into the soup with or without stirring.

Science Stuff Ahead

We used five treatments in determining the number of bacteria transferred from the mouth to chicken broth. Each treatment started with 80 mL of boiled, diluted chicken broth that we let cool to 170°F (77°C). The five treatments were (1) control, (2) dip, (3) dip–stir, (4) dip–mouth, and (5) dip–stir–mouth. After each treatment, 1 mL of the diluted broth was first directly pour-plated into plate count agar, and 1 mL of that was serially diluted in 0.1 percent peptone water and then pour-plated using PCA agar. Plates were incubated at $37°C \pm 2°C$ for 48 hours. Plates with serial dilutions yielding 25 to 250 colonies were counted and then converted to CFUs/20 mL of rinse.

1. **Just broth (control).** One mL of the diluted broth was sampled without inserting the sterile teaspoon in the broth, and bacterial populations were counted as described in Experiment 8-1.
2. **Dip the spoon.** Six spoonfuls of broth were removed from 80 mL of broth using a sterile spoon without inserting it into the mouth after removing each spoonful. One mL of broth was sampled and the bacterial population counted as described previously.
3. **Dip and stir with a spoon.** This treatment was the same as the dip-the-spoon treatment, except the broth was stirred three times while dipping.
4. **Dip the spoon after inserting it in the mouth.** Six spoonfuls of broth were removed from the 80 mL of broth using a sterile spoon that had been inserted in the mouth after removing each spoonful.
5. **Dip and stir the spoon after inserting it in the mouth.** The treat-

ment was the same as the dip–mouth
treatment except the spoon was stirred
three times while dipping.

Well, hello again!

Results of Experiment 8-2

If two people share a bowl of soup, is there a transfer to the soup of
bacteria that can be removed in the next spoonful? The answer is definitely yes. In fact, between 70,000 and 90,000 bacteria were trans-

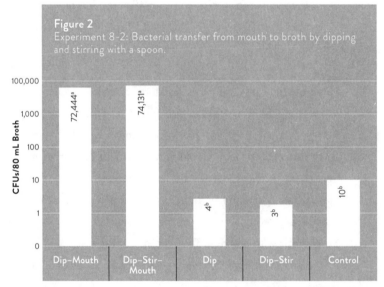

Figure 2
Experiment 8-2: Bacterial transfer from mouth to broth by dipping
and stirring with a spoon.

a,b Mean bacterial populations with different letters are significantly different at a level of 5%.
Ten participants.
CFU/80 mL Broth = total number of bacteria recovered from 80 mL of broth.
Dip–mouth = spoon dipped six times with the spoon inserted in the mouth after each dip.
Dip–stir–mouth = spoon was stirred three times before each of six spoonfuls were taken;
spoon was inserted in the mouth between spoonfuls.
Dip = dipping was performed six times in the broth without placing the spoon in the mouth.
Dip–stir = stirring was performed three times before six spoonfuls were taken without the spoon
inserted in the mouth after each dip.
Control = only the broth was sampled.

ferred to the broth after removing six spoonfuls of broth with a spoon that had been inserted in the mouth between each spoonful. There was no significant difference between dipping alone versus dipping and stirring with the mouth insertion treatments (Figure 2). It was estimated that about 10,000 bacteria were transferred each time the spoon was placed back into the broth after inserting a spoonful of broth into the mouth.

EXPERIMENT 8-3: Sharing Rice; Again, You Get More than Just Rice

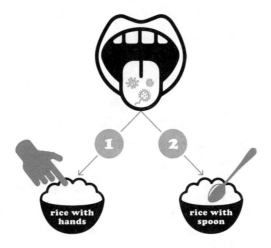

Materials and Methods

Would sharing rice be similar to sharing soup? In some cultures, people commonly use their hands to eat rice, so we tested hands and spoons along with the control treatments to compare how putting hands or spoons in your mouth beforehand transferred bacteria to the rice.

Science Stuff Ahead

Experiment 8-3 had two parts: (1) we tested bacterial transfer from the mouth to rice while being eaten with the hands; and (2) we tested bacterial transfer from the mouth to rice while being eaten with a spoon. Each participant was presented with 100 grams of cooked rice.

In part 1 of the experiment, we designed three treatments to determine the extent of bacteria transfer from mouth to rice using the hands or a spoon:

1. **Just rice (rice control).** Thirty grams of rice were aseptically placed in 100 mL of 0.1 percent peptone water without prior contact with the hand or mouth and then blended (Seward Stomacher 400 Circulator, Seward, Inc.) for 1 minute at 230 rpm. The blended sample was serially diluted in 0.1 percent peptone water and duplicate pour plates of the diluted samples prepared using PCA. The plates were incubated at 37°C ± 2°C for 48 hours. Plates from dilutions yielding 25 to 250 colony-forming units were counted and the counts converted to CFUs/gram of rice.

2. **Rice-hand.** Seven participants washed their hands with antibacterial soap for 20 seconds with warm water, rinsed with warm water for 10 seconds, and then towel-dried their hands using sterile paper towels. Using their hands, participants removed cooked rice from a sterile plate six times without placing their hands in their mouth. Bacterial count of the rice remaining in the bowl was determined as described for the rice control.

3. **Rice-mouth.** This treatment was the same as the rice–hand treat-

ment, except the participants inserted their hands into their mouths between each handful.

In part 2 of Experiment 8-3, we examined three treatments to determine the extent of bacteria transfer from mouth to rice using a spoon:

1. **Just rice (rice control).** This treatment was identical to the rice control from part 1 of the experiment.
2. **Rice-spoon.** Seven participants removed a spoonful of rice without putting the spoon in the mouth. They repeated this process six times.
3. **Rice-spoon-mouth.** This treatment was the same as the rice-spoon treatment except the participants inserted the spoon into their mouth between each spoonful.

Results of Experiment 8-3

Approximately 10,000 bacteria per gram of rice were recovered from the rice using either hands or spoons that had been inserted in the mouth. In comparison, less than 10 bacteria per gram were recovered from rice using either spoons or hands that had not been previously inserted in the mouth (Figure 3). When hands not previously placed in the mouth were used to sample the rice, the bacterial population in the bowl increased by nearly 100 bacteria per gram, compared to less than 10 bacteria per gram recovered from untouched rice. An average spoonful of rice weighed 9–10 grams, while the average handful weighed 11–12 grams. Thus, a spoonful or handful of

We need to stop meeting like this!

rice removed from a communal bowl shared with others could have ~100,000 more bacteria per spoonful or handful if others had previously taken samples with spoons or hands that had been placed in their mouths.

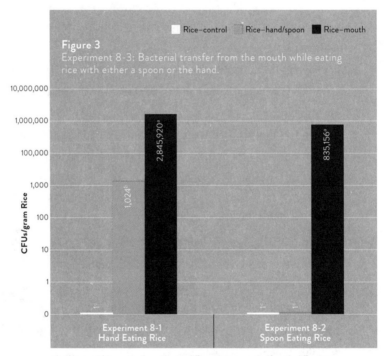

Figure 3
Experiment 8-3: Bacterial transfer from the mouth while eating rice with either a spoon or the hand.

a,b,c Means within an experiment having different letters are significantly different at a level of 5%. Seven participants.
CFUs/Gram rice: Total number of aerobic bacteria recovered from rice.
Rice–mouth = Rice was eaten with a spoon or hand that had previously been inserted in the mouth.
Rice–hand–no mouth: Five servings of rice were taken with a hand that had no contact with the mouth.
Rice–spoon–no mouth = Five servings of rice were taken with a spoon that had no contact with the mouth.
Rice–no mouth = Rice was sampled without contact with a spoon or hands.

EXPERIMENT 8-4: How Good Are the Methods of Recovering Bacteria?

Materials and Methods

You might wonder how well the method for recovering bacteria from the rice and soup works. Does it accurately represent the actual number of bacteria present in the rice or broth? The method of rinsing off bacteria from the hand or spoon or sampling a portion of the soup and then extrapolating the total number of bacteria in the whole bowl is at best an estimation. To test the accuracy of our microbial recovery procedures, we inoculated the rice and the soup with a known population of specific bacteria and then applied our recovery methods to see how well our findings matched the bacterial population that was used to inoculate the food.

Science Stuff Ahead

The effectiveness of our bacterial recovery procedures from diluted (1:4 ratio, as described in Experiment 8-2) Swanson chicken broth (Campbell Soup Company) and Mahatma extra-long-grain enriched rice (Riviana Foods Inc.) was determined by inoculating each food with an ampicillin-resistant strain of *E. coli*. Rice was cooked in a microwave oven (1,000 watt, Magic Chef) for 15 minutes (1:2 ratio of rice to water). All recovery methods were repeated three times and the results averaged.

Recovery Control A culture of ampicillin-resistant *E. coli* was grown overnight at 37°C ± 2°C in 10 mL of tryptic soy broth (Difco Laboratories) containing 100 parts per million (ppm) of ampicillin. The culture was centrifuged (International Equipment Company) at a speed of 1,000 times gravity (×g) for 15 minutes, the spent broth was discarded, and the bacterial cell pellet was resuspended in 10 mL of sterile 0.1 percent peptone water. One mL of this suspension was serially diluted in 0.1 percent peptone water, 0.1 mL of the dilutions were surface-plated onto tryptic soy agar plates, and then the plates were incubated at 37°C ± 2°C for 48 hours. Plates from dilutions having from 25 to 250 colonies were counted, and the population was converted to log CFUs/mL.

Recovery from Broth One milliliter of the washed suspension from the overnight ampicillin-resistant *E. coli* culture (prepared as described for recovery control) was inoculated into 80 mL of diluted (1:4) chicken broth. The inoculated broth was gently shaken for 20 seconds. One mL of the mixture was removed and serially diluted in 0.1 percent peptone water and then the bacterial population was enumerated as described for the recovery control samples. The percentage of recovery was calculated as follows:

$$\% \text{ recovery} = \frac{\text{number of cells recovered from broth}}{\text{number of cells recovered from control}} \times 100$$

Recovery from Rice One mL of the washed suspension from the overnight ampicillin-resistant *E. coli* culture (prepared as described for recovery control) was inoculated into 30 grams of cooked rice and thoroughly mixed. The inoculated rice was placed in 100 mL of 0.1 percent peptone water, blended (Seward Stomacher 400 Circulator, Seward, Inc.) for 1 minute at 230 rpm, and 1 mL of the blended mixture was sampled. The bacterial population was enumerated as described for the recovery from control samples. Percent recovery was calculated as follows:

$$\% \text{ recovery} = \frac{\text{number of cells recovered from rice}}{\text{number of cells recovered from control}} \times 100$$

Results of Experiment 8-4

Our recovery methods enabled us to recover about 98.6 percent of the bacteria that were inoculated into the chicken broth and about 89.7 percent of those inoculated into the rice. Not too bad. The higher recovery percentage observed in diluted broth compared to rice was likely due to the compositional (higher moisture) and physical (liquid vs. solid) differences between the foods. Chicken broth provides an enriched medium for the growth of the microorganism.[7] Furthermore, bacteria may have adhered to the rice grains to some extent and thus limited their complete recovery.

THINGS TO CONSIDER

Consubstantiation may spread friendship—and disease. Remember how psychologists said sharing food was an act of friendship and intimacy? Well, we also know that the oral cavity can be a source of pathogenic microorganisms.[8] At least five infectious diseases are known to transfer through oral saliva droplets and aerosols, and those include meningitis, pneumonic plague, tuberculosis, influenzas, Legionnaires' disease, and severe acute respiratory syndrome (SARS). If infectious agents are able to transfer from the oral cavity through the air, then bacteria and viruses likely can also be transferred between humans by exposure to utensils or foods that have become contaminated with saliva.[9]

The CDC recommends covering the mouth while sneezing and coughing to prevent spreading serious respiratory illnesses like influenza, respiratory syncytial virus (RSV), whooping cough, and SARS.[10] Additionally, contaminated surfaces can act as vectors for spreading diseases, since disease agents originate from the mucus of infected persons who have contaminated these surfaces.[11] Orally contaminated foods may also be transfer vectors, especially because infectious agents can survive outside the body for a long time—48 hours or more, depending on the specific agent and type of surface or food.

So, it turns out that people who share food and drinks are sharing more than good vibes. In fact, hands are a major cause of cross-contamination in food service, and oral bacteria left behind after someone takes a mouthful can later contaminate unsuspecting diners. Besides the intentional intimate sharing of food among friends, eating with a spoon or hands from a shared bowl is common when having snacks in many places such as taverns, entertainment venues, and parties.

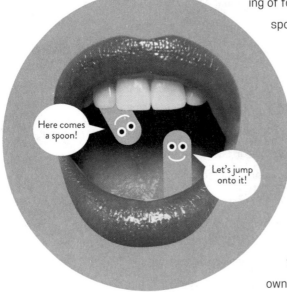

According to the World Health Organization, 27 percent of reported infections from foodborne pathogens are caused by person-to-person or person-to-food contamination.[12] Share food at your own risk. It's really a balance between intimacy and, well, sharing saliva. The first moment of eating an item together could be a turning point in a relationship. For example, it could be the moment when a casual acquaintance becomes your significant other. She offers you a bite of pastry from which she has just taken a large bite. Do you take the plunge and chomp down over the same end where she has just deposited her saliva? Do you turn down a sweet dessert because you're trying to keep the relationship in a holding pattern, hoping she doesn't get offended? Or, since you really want the pastry, do you take an even more definitive approach and essentially end the relationship by turning the pastry around and biting off the other end? These are the important decisions food sharing has imposed on us, and we must be ready to deal with them at any moment.

9 PASS THE POPCORN, PLEASE

Chapter 9

PASS THE POPCORN, PLEASE

Here we go, sharing food again. Sharing a jumbo tub of buttery popcorn at a movie theater is a common practice that may be done purely out of convenience, or it might have social implications (related to consubstantiation and intimacy). Whatever the reason, sharing popcorn with more than one person results in people putting their hands in their mouths and then back into the tub to grab more popcorn. We wondered, is bacteria being transferred to the popcorn from hands and mouths?

FIRST, A LITTLE POPCORN HISTORY

According to the Popcorn Board—and yes, there is such a thing—the oldest ears of popcorn ever found worldwide were in caves located in New

Mexico in 1948 and were dated at about 4,000 years old.[1] In 1519, Spanish conquistador Hernán Cortés described the first written encounter with popcorn, saying it was being consumed and used as decoration by the Aztecs in what is today Mexico. In the early 1800s, sailors coming to the United States from the Chilean seaport of Valparaiso introduced popcorn as a snack food.[2] It soon became a crowd-pleasing staple sold on city streets and at fairs, stadiums, and railroad stations in the United States. Its popularity really erupted in the United States during the Great Depression of the 1930s, when movie theater proprietors realized popcorn could be sold for a profit at a low cost. Movie theater popcorn continued to grow in popularity into the 1950s, until the advent of home televisions reduced theater attendance. Easy-to-use products such as Jiffy Pop and microwave bags have returned popcorn to its preeminent position as a snack food.

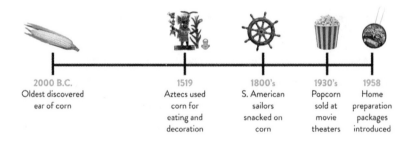

2000 B.C.	1519	1800's	1930's	1958
Oldest discovered ear of corn	Aztecs used corn for eating and decoration	S. American sailors snacked on corn	Popcorn sold at movie theaters	Home preparation packages introduced

DID YOU KNOW? NOT ALL CORN IS POPCORN

Many varieties of corn are grown, but there are only six general types:[3] Dent corn, the most common type, is mostly used for animal feed. Flint corn has similar uses as dent corn. Flour corn is used to produce cornmeal. Pod corn, which has multicolored kernels, is used mainly for decoration. Sweet corn has a soft exterior, moist interior, and high sugar content that make it good for eating off the cob. Popcorn has special characteristics, including a hard exterior and moist center so it will pop. Popcorn is the only type of corn that will successfully pop, and most of the world's popcorn is grown in nine U.S. states: Iowa, Illinois, Indiana, Kansas, Kentucky, Michigan, Missouri, Nebraska, and Ohio. When popcorn kernels are heated to around 400°F, the liquid moisture on the inside vaporizes into steam, creating over 100 pounds per square inch of pressure on the hard exterior hull. This causes the soft, fleshy inside to "pop" to the outside of the kernel, and voilà—popcorn![4]

HANDS, POPCORN, AND SURFACES—OH MY

Popcorn is a snack food that is often consumed by more than one person from one large container, whether it's a large bowl, a box, or a bag. What's more, when consuming popcorn, people are likely to place their hands into a container multiple times after they have placed their hands in their mouth—we know, it's pretty hard to eat just one handful of the stuff.

To find out how much interest there is in hand hygiene, we searched PubMed, an online database of published biomedical papers, for the keyword *handwashing*. This search yielded 932 papers for the years 1985–1994, 3,538 papers for 1995–2004, and 5,463 papers for 2005–2014.

As we've repeatedly shown in this book, hands play a major role in transmitting infection among humans in public settings, including via food.[5] Furthermore, 28 percent of commuters in five UK cities were found to carry

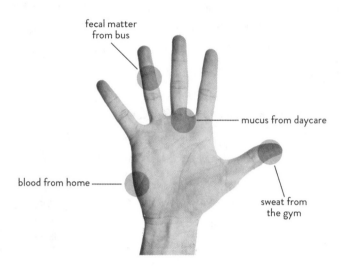

fecal matter
from bus

mucus from daycare

blood from home

sweat from
the gym

fecal bacteria on their hands, and interestingly, bus commuters had higher bacterial counts than train riders.[6] Environmental surfaces can readily transmit bacteria to hands when touched, and while most of these bacterial and viral species are harmless, some pose a potential health risk and can be transmitted to food or other surfaces by contact with hands. Since bacteria can be picked up by hands when eating popcorn in public places such as movie theaters, we set forth to determine the average number of bacteria transferred from hands to popcorn while eating popcorn.[7]

Let's first ask ourselves, what's a handful? For consistent data collection, we first had to determine the amount of popcorn in a typical handful. So we kept filling a test bowl with 5 grams of popcorn and then asking each of 11 participants (replicated three times for a total 33 observations) to remove a handful from the bowl. For each observation, we separately weighed the amount of popcorn removed and that remaining in the bowl. The averages, medians, minimums, and maximums for the amount of popcorn removed per handful and the amount remaining in the bowl were calculated (Figure 1). So, depending on the size of your hand—and how hungry you are—we estimate that you are getting between 1.7 and 3.4 grams of popcorn per handful. That's about three handfuls per cup. Three handfuls of air-popped corn have about 31 calories, 6 grams of carbohydrate, a negligible amount of fat, 1 gram of protein, and 1 gram of fiber—it's actually a nutritious snack!

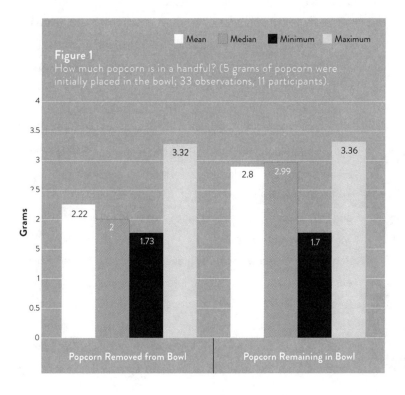

Figure 1
How much popcorn is in a handful? (5 grams of popcorn were initially placed in the bowl; 33 observations, 11 participants).

■ Mean ■ Median ■ Minimum ■ Maximum

Popcorn Removed from Bowl: Mean 2.22, Median 2, Minimum 1.73, Maximum 3.32
Popcorn Remaining in Bowl: Mean 2.8, Median 2.99, Minimum 1.7, Maximum 3.36

Grams

EXPERIMENT 9-1: Popcorn with a Side of *E. coli*

Materials and Methods

After estimating the amount of popcorn in a handful, we placed 5 grams of popcorn in a bowl and used 2.5 grams to represent a handful of popcorn. Our nonpathogenic marker strain of *E. coli*, which contains a fluorescent gene, was employed to track the transfer and survival of bacteria on popcorn. We used Orville Redenbacher's Naturals brand of microwave popcorn for this study, and we made sure that the salt and butter concentrations in the popcorn would not inhibit the recovery of the inoculated *E. coli* test strain. To determine if sharing popcorn from a common container contributes to cross-contamination, we tested these five treatments:

1. **Inoculated hand.** After they washed their hands with soap and warm water, we inoculated each participant's hands with 1 mL of the *E. coli* preparation and then asked them to rub their hands together for 30 seconds. Next, participants inserted their dominant hand into a sterile bag containing 20 mL of sterile 0.1 percent peptone water and then massaged the hand for 30 seconds to recover bacteria washed from their hands.

2. **Popcorn removed with sterile gloves.** After washing their hands, each participant put on sterile latex-free gloves. Five grams of freshly popped popcorn was aseptically weighed into a sterile bowl. The participants then removed approximately 2.5 grams of popcorn from the bowl using their one dominant, gloved hand. The handful removed from the bowl was placed into a sterile bag containing

20 mL of sterile peptone water and then mixed for 30 seconds to recover bacteria from the popcorn.

3. **Popcorn remaining in the bowl after removing popcorn with one sterile glove.** The bacteria on popcorn remaining in the bowl from the second treatment were recovered the same way as in treatment 2.

4. **Popcorn removed with inoculated hands.** After hand washing, we inoculated each participant's hands with 1 mL of the *E. coli* preparation and asked them to rub their hands together for 30 seconds. Participants then removed approximately 2.5 grams of popcorn with their dominant hand from a sterile bowl containing 5 grams of popcorn. Bacteria transferred to the popcorn were recovered as in earlier treatments.

5. **Popcorn remaining in the bowl after removing a handful with an inoculated hand.** After participants had removed approximately 2.5 grams of popcorn with an inoculated hand, the bacteria transferred to the popcorn remaining in the bowl were recovered as described earlier.

Science Stuff Ahead

Bacteria were recovered from the 20 mL of rinse solution from each treatment by first serially diluting 1 mL of the rinse solution in 0.1 percent peptone water. Then 0.1 mL of each dilution was surface-plated onto tryptic soy agar plates. The plates were incubated at 37°C for 24 hours. Then we counted the plates from dilutions having 25–250 colonies and calculated the number of colony-forming units (CFUs) per hand or per gram of popcorn. The percentage of *E. coli* transferred from inoculated hands to popcorn was calculated as follows:

$$\% \textit{ E. coli} \text{ transferred} = \frac{\text{number of } \textit{E. coli} \text{ on popcorn}}{\text{number of } \textit{E. coli} \text{ on hands}} \times 100$$

The experiment was replicated 17 times with 8 observations per replicate, resulting in 136 total observations. Descriptive statistics (mean and standard deviation) were determined using Statistical Analysis Software (SAS).[8]

Results of Experiment 9-1

First of all, when popcorn was handled with gloved hands, the *E. coli* population was below the level of detection (< 10 bacteria), as expected. For popcorn handled with hands inoculated with *E. coli*, 82 percent (112 popcorn samples out of 136) were *E. coli*–positive. For

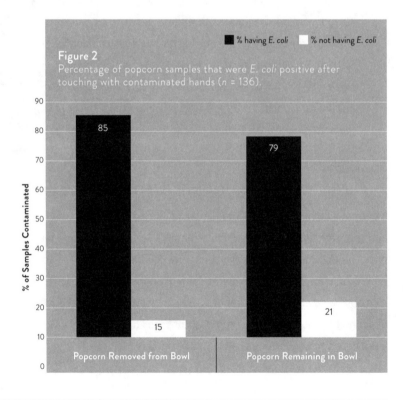

Figure 2
Percentage of popcorn samples that were *E. coli* positive after touching with contaminated hands (*n* = 136).

■ % having *E. coli*   ☐ % not having *E. coli*

the handful of popcorn removed, 85 percent of the samples had some *E. coli* while 79 percent of the popcorn remaining in the bowl was *E. coli* positive (Figure 2). These findings clearly demonstrated that *E. coli* does transfer from inoculated hands to popcorn by the simple action of grabbing a handful.

A typical serving size (based on package labels) of popcorn generally ranges from 28 to 35 grams. Thus, one serving would generally require the consumer to reach into the bag about twelve times, thereby potentially increasing the population of bacteria transferred to the popcorn by twelvefold. Most popcorn packages contain 2–4 servings, which would

require 24–48 handfuls. When we eat popcorn in public places like movie theaters, our hands pick up bacteria from plenty of surfaces. As previously stated, one in four public surfaces were found to be highly contaminated, including popcorn-filled movie theaters. Additionally, bacteria can survive on common public surfaces for weeks to months, making nearly all public places reservoirs for bacteria. The bacteria can be transferred to surfaces not only by direct hand contact but also by sneezing, coughing, speaking, and even breathing.[9]

When only popcorn samples containing *E. coli* were included in our calculations, the average population of *E. coli* recovered per handful of popcorn (2.5 grams) was 185 CFUs while the average for popcorn remaining in the bowl was 48 CFUs per handful (Figure 3). An average

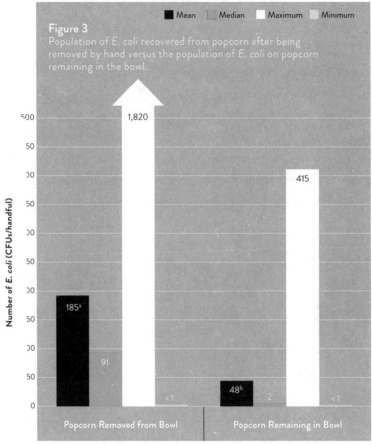

Figure 3
Population of *E. coli* recovered from popcorn after being removed by hand versus the population of *E. coli* on popcorn remaining in the bowl.

Mean ░ Median ▢ Maximum ▨ Minimum

Number of *E. coli* (CFUs/handful)

500

50

00

50

00

50

00

50

0

1,820

415

185[a]

91

< 1

48[b]

2

< 1

Popcorn Removed from Bowl

Popcorn Remaining in Bowl

a,b Popcorn taken and remaining using inoculated hands having the same letter are not significantly different at a significance level of 5%; *n* = 136.

of 500,000 *E. coli* were found on inoculated hands, so the percentage of bacteria transferred to popcorn was relatively low (0.2 percent maximum transfer) for popcorn taken and popcorn remaining (0.009 percent maximum transfer) when compared to the 11 percent average transfer we detected for hands to menus (Chapter 3). However, the maximum number of *E. coli* transferred was 1,820 cells per handful of popcorn removed and 415 cells per handful of popcorn remaining in the bowl. Considering the number of handfuls required to eat a large bag of popcorn (4 cups or 4 servings, based on package labels), a maximum of 87,000 *E. coli* cells could be transferred by hand to the popcorn, and at most 20,000 *E. coli* cells would remain in the bowl. By the time you're scraping the bottom of the bag, you might as well be kissing the person sharing the bag with you.

THINGS TO CONSIDER

In 1990, two researchers named Elizabeth Scott and Sally Bloomfield found that hand contact with contaminated surfaces plays an important role in the transmission of bacteria.[10] Other investigators have reported that transmission of diseases by human hands to common surfaces touched by others is a key public health concern.[11] As we now know, sharing popcorn results in the transfer of bacteria to the popcorn. Moreover, eating popcorn in public venues could result in "self-contamination" by touching surfaces such as doors, seats, handrails, commodes, and armrests. Given this, we can say that eating popcorn in public places diminishes the consumers' likelihood of having "clean" hands.

Another consideration is that after placing your fingers in your mouth when eating popcorn, you might be transferring disease-causing agents from your mouth to your hands. From 500 to 700 types of bacterial species are found in the mouth.[12] Moreover, bacterial populations in the mouth are generally very high.[13] According to the Centers for Disease Control and Prevention (CDC), the mouth is a prime source for spreading infectious disease.[14] In 2004, Steven Harrel and John Molinari identified five different infectious diseases that are spread by saliva; in 2007, researchers from the Japanese National Center for Geriatrics and Gerontology found that some of these bacteria can be life threatening.[15] Most healthy adults can tolerate exposure to infectious agents, but sharing of popcorn by those with weakened immune systems may create a significant risk of illness from low-level contamination.

When the ABC television show *20/20* did a random sampling of movie theaters in New York and Los Angeles, the seats and armrests were found to be highly contaminated with fecal bacteria, among various other microbes.[16] This may not surprise you, given what you've learned from this book. The scientist involved in the study, Philip Tierno, said, "One of the things, when I'm eating popcorn, is not to touch the seats with your right hand. If you're going to hit the armrest, keep your hand in the air . . . always have a good hand and a bad hand."

So besides finding other people's "mouth" bacteria in the bowl of popcorn, you could also be acquiring microbes from armrests, handrails, seat cushions, commodes, and other contact surfaces at your local entertainment venue. How often do you suppose those seats and armrests are cleaned in your local theaters? Our guess is not often enough.

As if things weren't bad enough, there's also something called popcorn lung. This serious lung disease can result from inhaling added flavoring compounds that become volatile when heating popcorn as well as other foods.[17] Although it's no health threat for the average consumer, smokers and e-cigarette users are at risk. But the main concern is for workers at popcorn and other food-manufacturing locations that use flavorings, specifically diacetyl, a compound found in butter flavoring. Diacetyl also occurs naturally in fermented products such as sour cream, buttermilk, and alcoholic beverages.

So the next time you're at the movies, you might want to consider what the micro-biologists say—and be careful who you share that popcorn with. Food and socializing are closely linked, but we all might want to be well aware of the potential to pick up microbes in public places. Speaking of socializing, in the next chapter we'll consider why so many of us are annoyed by those who practice the dreaded double-dip.

DIP CHIPS AND DOUBLE-DIPPING

Chapter 10

- -

DIP CHIPS AND DOUBLE-DIPPING

Let's go back to Chef Alton Brown's comment in the opening of Part 3. There, we considered his suggestion that a dip is defined based on its ability to "maintain contact with its transport mechanism over three feet of white carpet." Wise words, Chef Brown. This definition depends on how well a dip clings to a chip or cracker as well as the dexterity of the dipper and the shape of the chip.

Now picture this: While mingling at a social event, you notice someone performing the infamous double-dip. This means that they dipped a partially eaten chip or cracker back into the dip. How do you react? Do you step away from the dip and have to revise your opinion of the double-dipper? Or do you just shrug and help yourself to some of it? Double-dipping is an act that many people consider disgusting, regardless of the "transport mechanism" or dip. And yet others see no danger in double-dipping and feel no shame in dunking the same chip two, three, or more times into the dip.

Whether the act is crude, rude, or unrefined, the real question remains: Does the double-dip transmit bacteria from the double-dipper's mouth to the dip? Does this practice pose any health concerns for those who consume a dip after a double-dipper? How many bacteria are transferred from the perp's mouth to the dip and then survive to be consumed by the next person to scoop up the now-contaminated dip?

These are vital questions, and we're grateful to *Seinfeld* for bringing the practice of double-dipping into the popular imagination during its season 4 episode "The Implant." The classic scene, if you haven't viewed it, is immortalized on YouTube with a 1-minute interchange between George (Jason Alexander) and Timmy (Kieran Mulroney) at the buffet table during a wake for a relative of George's girlfriend.

Sorry, Timmy. But I don't dip that way.

As scientists, we decided to determine in the lab whether double-dipping is like "putting your whole mouth in the dip," as Timmy says, and if bacteria are transferred to a dip by this practice. We set up a series of three experiments with eight dippers, three dips, and one transport mechanism: crackers.

Before conducting any of the dipping experiments, we needed to estimate the number of bacteria in each participant's mouth. Lucky for us and our experiments, human oral bacteria are easy to find because they are not naturally found anywhere else in the body. Previous studies estimated that between 300 and 1,000 different types of bacterial species are found in the human mouth. But what population of bacteria did this amount to?

We obtained estimates of the bacterial count in our eight participants' mouths by having each of them rinse with 20 milliliters (mL) of sterile water. We collected the mouth rinse, cultured it, and then calculated
the number of bacteria in the rinse water. There were, on average, over 1
million bacteria per milliliter, ranging from 575,000 to over 4.8
million cells. Remember, this is the number of bacteria in 1 mL of
rinse. The number swished out of the mouth was at least twenty
times that much. In reality, total number of bacteria per mouth
has been estimated at 20 billion. Now that is truly a mouthful.

EXPERIMENT 10-1: Does the "Transport Mechanism" Carry More than Just Dip?

Materials and Methods

For Experiment 10-1, our goal was to compare bacterial populations
of two sterile solutions after one was exposed to bitten crackers and
the other was exposed to unbitten crackers. Low-sodium Wheat Thins
(Nabisco) were used for dipping in all three experiments. In every case,
the bitten or unbitten portion of the cracker was dipped into sterile water
for 3 seconds and then the cracker was discarded. For all dipping experiments, the crackers were held with sterile gloves and an equal number
of tests were done with bitten and unbitten crackers.

To simulate double-dipping, a dipper would bite a new cracker, dip
the remaining portion into the 20-mL solution, discard the cracker, and
repeat for a total of three bites and dips. Now this may sound like triple-
dipping, but we were attempting to simulate a party environment. In this
setting we theorized that a double-dipper would probably double-dip each

cracker and consume multiple crackers during a single visit to the dip. Picture George Costanza at the dip bowl, dipping one chip after another until he's had enough. (Because seriously, who can stop after just one chip?) Since we did not expect much bacteria to be transferred, we also included a six-dip, or "worst-case," scenario. We tested the rinse water after three dips and then again after six dips. Each dipper followed the identical process with unbitten Wheat Thins. Data for experiments 10-1, 10-2, and 10-3 were analyzed for differences at the 5 percent significance level using the Statistical Analysis System (SAS).[1]

Results of Experiment 10-1

Damn you, Timmy! Biting crackers before dipping transferred a significant number of bacteria into sterile water, resulting in over 1,800 bacteria per mL of water as compared to only 5 bacteria per mL of water exposed to three crackers that were not bitten before dipping (Figure 1). The unbitten crackers dipped six times, on the other hand, transferred less than 10 bacteria per mL (0 to 21 range), while sterile water exposed to six bitten crackers had more than 2,500 bacteria per mL. This is indisputable evidence that oral bacteria were, in fact, transferred to the water by biting a cracker before dipping. Now you know.

EXPERIMENT 10-2: Tart Stuff, but Not Tart Enough

Materials and Methods

We also wanted to evaluate whether the acidity of the dips would protect us by killing oral bacteria once they were introduced into the dip. To

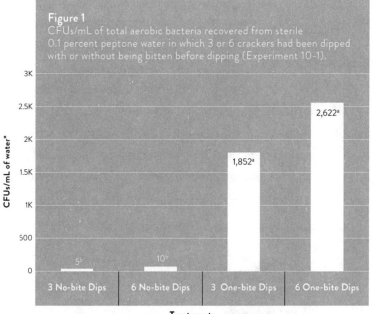

Figure 1
CFUs/mL of total aerobic bacteria recovered from sterile
0.1 percent peptone water in which 3 or 6 crackers had been dipped
with or without being bitten before dipping (Experiment 10-1).

3 no-bite dips = 3 different wheat crackers, each dipped once without biting
6 no-bite dips = 6 different wheat crackers, each dipped once without biting
3 one-bite dips = 3 different wheat crackers, each bitten once before dipping
6 one-bite dips = 6 different wheat crackers, each dipped once without biting
a,b Means with different letters are significantly different at a level of 5%.
Standard error = 0.1, n = 16.
*CFUs/mL of water = colony-forming units per milliliter of dipping solution.

test this possibility, we evaluated the effect of dipping crackers in sterile water that was adjusted to pH levels of 4, 5, and 6. (As previously mentioned, 7 is neutral, less than 7 is acidic, and more than 7 is alkaline). These pH levels were based on the average pH of samples tested from commercial dips of salsa (pH 4.0), chocolate (pH 5.1), and cheese (pH 6.1).

The three pH solutions were sampled for bacteria immediately after dipping from both bitten and unbitten crackers. We also tested the dips for bacteria 2 hours later, to simulate "party" conditions and see if the longer exposure to acid would kill any bacteria remaining in the dip from the double-dipper.

Results of Experiment 10-2

In Experiment 10-2, we tested whether the acid level and holding time after dipping affected the bacteria. Some typical foods and items having various pH levels are shown in Figure 2. This experiment supported the findings from Experiment 10-1 in that biting a cracker before dipping transferred oral bacteria to the dipping solution, but we also found that

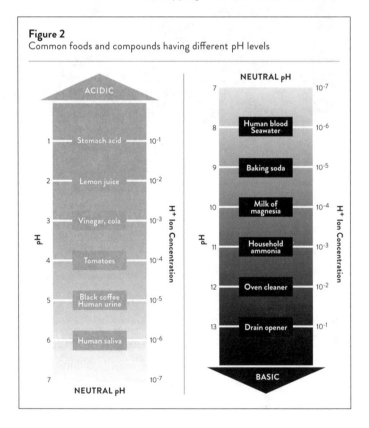

Figure 2
Common foods and compounds having different pH levels

ACIDIC

NEUTRAL pH

pH		H⁺ Ion Concentration

1 — Stomach acid — 10^{-1}

2 — Lemon juice — 10^{-2}

3 — Vinegar, cola — 10^{-3}

4 — Tomatoes — 10^{-4}

5 — Black coffee / Human urine — 10^{-5}

6 — Human saliva — 10^{-6}

7 — 10^{-7}

NEUTRAL pH

7 — 10^{-7}

8 — Human blood / Seawater — 10^{-6}

9 — Baking soda — 10^{-5}

10 — Milk of magnesia — 10^{-4}

11 — Household ammonia — 10^{-3}

12 — Oven cleaner — 10^{-2}

13 — Drain opener — 10^{-1}

BASIC

lower pH levels (more acidity, at pH 4 and 5) reduced the population of bacteria detected in the dipping solution after being held for 2 hours at room temperature (Figure 3). But let's not get too excited. The bacterial populations were reduced from around one to four thousand to around one hundred to two thousand, which sounds impressive but is not low enough to eliminate harmful microbes possibly added to the dip by the dipper. As we will find out later, the pH of a dip is less bactericidal in the real world of dips.

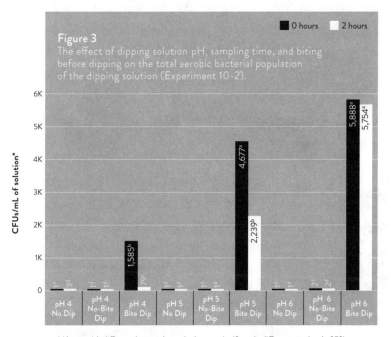

Figure 3
The effect of dipping solution pH, sampling time, and biting before dipping on the total aerobic bacterial population of the dipping solution (Experiment 10-2).

a–d Means with different letters above the bar are significantly different at a level of 5%; n = 18 (nine participants repeated the experiment twice for a total of 18 observations for each treatment).
Standard error = 0.1.
No dip = no cracker was dipped in the water.
No-bite dip = cracker was not bitten before dipping in the water.
Bite dip = cracker was bitten before dipping in the water.
*CFUs/mL of solution = colony-forming units per milliliter of dipping solution.
0 hours = solutions were sampled immediately after dipping at 0 hours.
2 hours = solutions were sampled after being held at room temperature for 2 hours.

EXPERIMENT 10-3: The Money Shots

Materials and Methods

In the third experiment, we put the myth to the test by double-dipping into three different—and delicious—dips. We chose them for their popularity, their viscosity, and their likelihood of being double-dipped. The dips we used were All Natural Tostitos Chunky Hot Salsa (Tostitos brand, Frito-Lay), pH 4; Genuine Chocolate Flavor Hershey's Syrup (Hershey Corp.), pH 5.1; and Fritos Mild Cheddar Flavor Cheese Dip (Frito-Lay), pH 6.1.

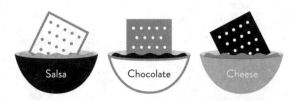
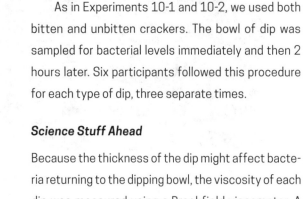

Salsa   Chocolate   Cheese

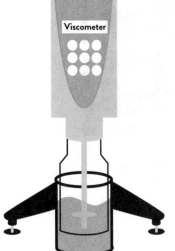

As in Experiments 10-1 and 10-2, we used both bitten and unbitten crackers. The bowl of dip was sampled for bacterial levels immediately and then 2 hours later. Six participants followed this procedure for each type of dip, three separate times.

Science Stuff Ahead

Because the thickness of the dip might affect bacteria returning to the dipping bowl, the viscosity of each dip was measured using a Brookfield viscometer. A Brookfield viscometer is an instrument that determines the resistance of fluids to flow (viscosity) by using a spinning spindle placed into the fluid. The viscometer motor rotates the spindle at a set speed, and the viscometer measures the resistance to rotation and then determines viscosity.

The amount of dip adhering to the cracker was estimated by measuring the weight loss from the bowl of dip after three dips. After each dip, we allowed the dip to drain from the cracker and back into the dip bowls for 5 seconds.

All bacterial plating was performed in duplicate. We then investigated the effects of dipping treatments, pH levels, hold times, and dip types on dip bacterial populations. A statistical analysis of these treatment effects was performed at significance level of 5 percent using the Statistical Analysis System (SAS).[1]

Results of Experiment 10-3

HEADLINE: Double-Dipping Is Like Putting Your Mouth in the Dip. When the salsa, chocolate, and cheese dips were double-dipped, from 100 to 1,000 more bacteria ended up in the dip when compared to dipping without first nibbling (Figure 4). This is sweet, sweet scientific proof for most of us. And a devastating setback for a few of us. You know who you are.

Let's have the good news first: dipping an unbitten cracker did not add a significant number of bacteria to the dip. The bad news is that double-dipping in salsa, one of the most popular dips around, initially added more bacteria to the dip as compared to those containing chocolate or cheese. It was determined that unlike the more viscous chocolate and cheese dips, the lower viscosity of the salsa resulted

in more of the salsa dripping back into the bowl of dip. The viscosity of the dipping solution—and more specifically, how well the dip adhered to the cracker—was inversely related to the transfer of bacteria to the dip. In other words, the more dip that drops back into the dipping bowl, the greater the number of bacteria transferred from the bitten cracker back into the dip. In this sense, salsa may not pass Chef Alton Brown's dip definition of being able to maintain contact with its transport mechanism over three feet of white carpet.

There's more good news: salsa killed off about 90 percent of the bac-

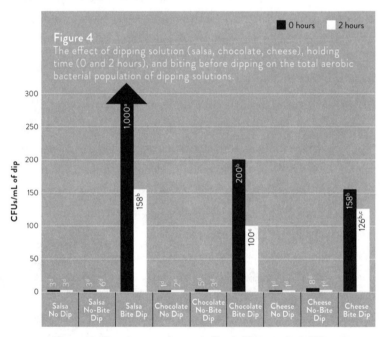

Figure 4
The effect of dipping solution (salsa, chocolate, cheese), holding time (0 and 2 hours), and biting before dipping on the total aerobic bacterial population of dipping solutions.

a–d Means with different letters over bars are significantly different at a level of 5%; n = 12.
Standard error = 0.1.
No dip = no cracker was dipped in the sauce.
No-bite dip = cracker was not bitten before dipping in the sauce.
Bite dip = cracker was bitten before dipping in the sauce.

teria after 2 hours. But we also have bad news: plenty of bacteria were left in the salsa after 2 hours—about the same as in the chocolate and cheese, which didn't kill as many of the bacteria during the two-hour holding time. This finding was assumed to be due to the lower pH of the salsa dip or possibly the presence of other antimicrobial properties in the dip.

The revolting (in our humble opinion) practice of double-dipping may have more relevance to fellow dippers than just being a "yuck" factor, since disease-causing microbes can exist in the mouth. In fact, complex statistical models have been developed to predict the spread of disease from a single individual to large populations. Humans harbor and spread disease by sneezing, handshaking, coughing, and possibly double-dipping. Flash back to the nineteenth century and the infamous cook Mary Mallon, or "Typhoid Mary," who spread typhoid (*Salmonella Typhi*) to many families by handling their food. Individuals sharing a dip may be passing disease-causing agents to the dip, yet not show outward clinical signs of illness. Bacteria located in the mouth often attach to each other and to mouth surfaces to form dynamic bacterial communities.

Dips are often high-acid foods (pH < 4.6) that kill or inhibit the growth of many types of bacteria. In fact, in our study, the low-pH salsa dip did have an inhibitory effect on bacterial populations when the dip was held for 2 hours; however, the bacterial populations were still much higher than those in dips that had not been double-dipped. Bacteria can also modify their environment to facilitate survival and growth, and repeated dilution and inoculation of a dip through double-dipping may further compromise the safety of that dip. What's more, dips also contain proteins, lipids, and other food components that can act as pH buffers to protect microorganisms.

On the other hand, probiotic bacteria (ingested microorganisms that have positive health effects) have been shown to survive and in some cases increase in population when added to an acidic cheese-based dip.[2] We touched on this finding in the introduction to Part 3, mentioning that "good" bacteria can also strengthen your immune system.[3]

THINGS TO CONSIDER

Did you know that you can send a stool sample to the American Gut project[4] to have it analyzed? The American Gut is a citizen science group that is funded by individual donations. You can get your own microbiome sequenced for $89. Some other groups studying the bacteria in your gut include the Human Microbiome, Global Gut, and Personal Genome projects.

Research into what affects the bacteria in our gut and how these bacteria affect our health is an expanding field. In fact, study of the human gut microbiome has a new area called psychobiotics that looks into how the microorganisms in our intestine can even affect our brain.[5] On the list of things that are strange but true, you can receive a fecal transplant from a donor with healthy poop. This is actually a radical but standard treatment for people who have a *Clostridium difficile* infection, since this pathogen is nearly impossible to eliminate with antibiotics.[6]

Fecal transplants are typically limited to people afflicted with a *C. difficile* infection, but one woman named Lauren Petersen took matters into her own hands after battling chronic fatigue brought on by years of antibiotic treatments for Lyme disease. She used an enema kit to perform her own fecal transplant from a healthy donor.[7] She appears to have made a complete recovery, is feeling very healthy, and is now a researcher in the field. Another item on the list of things you might not want to know about the human microbiome is "pooperoni." With the idea of using fermented meats as a way to ingest "good" bacteria,

"Good" bacteria delivery?

researchers from Spain isolated three bacteria from infant feces and used them to make sausage.[8] But enough fun facts about the colon; let's return to our exploration of double-dipping.

Getting back to the oral cavity, we want to remind you that the concern over human saliva as a way to spread infection is not a new concept. Dentist's offices and laboratories are a source of bacterial infection for patients and dental workers.[9] Water used for gargling and rinsing and dental rinse-water lines have been found to contain bacterial contamination originating from human saliva.[10] The spread of contaminating microorganisms from saliva by any means is a potential threat to human health and food safety.

Another interesting tidbit gleaned from our double-dipping study was that the thickness of the dip affected the amount of dip adhering to the cracker, which in turn determined how much contaminated dip fell back into the bowl. As a result, the viscosity of the dip had more effect than did the pH of the dip on the bacteria left in the bowl by double-dipping. In Experiment 10-2, where the viscosities of all dipping solutions were equal, the lower pH solution yielded the lowest initial bacterial population. But in Experiment 10-3, where viscosity and adhesion differed across dipping solutions, the dipping sauce with the lowest pH yielded the highest initial bacterial populations. Thus the physical properties of a dipping solution (viscosity, cohesion, and adhesion) also significantly affect the transfer of bacteria from a contaminated cracker to the dip.

It's debatable whether double-dipping poses a health risk. That depends on many factors, among them the type of chip or cracker, the population and types of microorganisms transferred, the number of double-dips, the type of dip, the health of the person who double-dipped, and the health of the people consuming the contaminated dip. Double-dippers are here to stay, so what can be done? Should we serve dip in individual dipping bowls, or offer disposable cups for each dipper? Or perhaps a thicker dip that keeps more on the chip and drops less from your lip back into the dip! These are merely suggestions, of course. We wouldn't want to stop the party.

Who am I to judge?

EPILOGUE

Food Microbes
and Safety

Epilogue

- - - - - - - - - - - - - - - - - -

FOOD MICROBES AND SAFETY

According to the Centers for Disease Control and Prevention (CDC), foodborne disease is certainly a problem. Here are the latest numbers we've found. Foodborne diseases cause an estimated 48 million illnesses and 3,000 deaths each year in the United States.[1] Of the 7,998 outbreaks (an outbreak can have hundreds of individual cases) with known causes, 3,633 (45 percent) were caused by viruses, 3,613 (45 percent) by bacteria, 685 (5 percent) by chemical and toxic agents, and 67 (1 percent) by parasites.[2] What's more, estimates place the annual costs of foodborne illness in the range of $55 to $77 billion.[3]

Although the food industry and federal and state regulatory agencies continue to explore ways to reduce foodborne illness outbreaks, consumers such as yourself can take steps to avoid food contamina-

I love when you forget me in the trunk!

tion. While many microbial control measures have been established, there is no way to destroy or remove all pathogenic and spoilage microorganisms from our food systems to ensure zero risk. Furthermore, public health experts place much of the blame for foodborne illness on mistakes made in the home. Blunders made in shopping, transporting, storing, preparing, or serving food can allow pathogens to survive and grow. Informed consumers—including those who have finished reading this book—can help extend the actions taken by industry and government by becoming more educated on how to buy, prepare, and store food safely. The key is to identify and understand the food safety risks associated with our food-handling practices and to decrease those risks to acceptable levels.

There is a solution to the food safety problem, and we can greatly reduce our risk by following some general rules to prevent foodborne illness in the home. It is essential that consumers think about food safety at each step, from shopping to cooking to storing leftovers. Information presented here is from several sources, including *Food Safety: A Consumer's Guide to Understanding*;[4] the U.S. Food and Drug Administration (FDA) and its Fight BAC Program;[5] and the Center for Food Safety and Applied Nutrition (CFSAN).[6]

Researchers at the CDC have identified five practices and behaviors that have contributed to most outbreaks of foodborne illness. They include poor personal hygiene, inadequate cooking, improper holding, contaminated equipment, and cross-contamination.[7]

We have pared down these five practices to the following eight action items.

1. Shop smart.
   - Take food straight home and refrigerate it promptly.
   - Don't buy anything you won't consume before the use-by or sell-by date.
   - Buy perishable foods last, and take them straight home to the refrigerator.

2. Chill: Refrigerate promptly.
   - Refrigerate or freeze perishables, ready-to-eat foods (those that you do not heat before eating), and leftovers within 2 hours of purchasing or preparation. Make sure the refrigerator is set no higher than 40°F and the freezer is set at 0°F.
   - Freeze fresh meat, poultry, or fish immediately if you can't use it within a few days (that is, 2 or 3 days).
   - Put packages of raw meat, poultry, or fish in a shallow pan or other container before refrigerating so juices won't drip onto other food.
   - If possible, leave a product in its store wrap; if a package is too large, divide the contents into smaller portions and then wrap and freeze what you don't plan to cook right away.

3. Clean: Wash hands and surfaces often.
   - Wash your hands with hot soapy water before and after preparing food. It is recommended to dry hands with

either disposable paper towels or a clean cloth that has not been used on food or food preparation surfaces. Be sure to wash your hands after using the bathroom, changing diapers, and playing with pets. The CDC recommends the following procedure for washing hands:[8]

1. **Wet** your hands with clean, running water (warm or cold), turn off the tap, and apply soap.

2. **Lather** your hands by rubbing them together with the soap. Be sure to lather the backs of your hands, between your fingers, and under your nails.

3. **Scrub** your hands for at least 20 seconds. Need a timer? Hum the "Happy Birthday" song from beginning to end twice.

4. **Rinse** your hands well under clean, running water.

5. **Dry** your hands using a clean towel or air-dry them.

- Wash kitchen towels often in the hot cycle of your washing machine; avoid using sponges, or put them in the dishwasher daily to kill bacteria.

- Wash your cutting boards, dishes, utensils, and countertops with hot, soapy water and a hot-water rinse after preparing each food item and before you go on to the next food item. Health experts recommend that before washing items with hot soapy water, you use warm water to flush away large food particles, or scrape them from the utensil or food contact surface. Follow this step by washing in a hot, soapy detergent solution and then thoroughly rinsing with clean, hot water to remove the detergent. Although sanitizers are not generally used in the home (except in automatic dishwashers), commercial food-handling operations use them to clean the surfaces that food contacts. Cleaning does remove soil and residues from surfaces, but sanitizing involves treating previously cleaned food contact surfaces with a chemical sanitizer or super-hot water (just under 200°F) to reduce the number of disease-causing microorganisms to safe levels. Sanitizing does not translate to complete sterilization, because some bacterial spores and a few highly resistant bacterial cells can survive. Depending on the type of dishwashing system used (manual vs. mechanical), hot water ranging from 171°F (dipping for 30 seconds) to 194°F is recommended for sanitizing utensils that have been in contact with food.[9]

4. Separate: Avoid cross-contamination.
   - Cut vegetables or salad ingredients first, then raw meat and poultry. Rinse fresh fruits and vegetables under running water.
   - Wash cutting boards, utensils, and countertops with hot, soapy water after cutting raw meat and poultry products and before slicing vegetable or salad ingredients.
   - Keep raw meat, poultry, eggs, and seafood and their juices away from ready-to-eat foods.
   - Never place cooked food on a plate that previously held raw meat, poultry, eggs, or seafood unless the plate has been thoroughly cleaned between uses.
   - Do not use a sponge to soak up meat and poultry juices. Use disposable towels.

5. Cook to proper temperatures.[10]
   - Thaw food in the refrigerator or microwave, not on the kitchen counter; marinate food in the refrigerator.

- Use a clean meat thermometer to measure the internal temperature of cooked foods to make sure meat, poultry, casseroles, and other foods are cooked all the way through (insert the thermometer into the approximate center of the product; for meats with bones, insert the thermometer into the center of the muscle and not directly next to a bone, which conducts heat faster than muscle).

- Cook ground beef, including meat loaf, to at least 160°F. At this temperature there is usually no pink left in the middle. Remember that ground beef is made by grinding together (mixing) different muscles, perhaps even from different animals, and therefore any microbial surface contamination will be mixed throughout the product. Cook ground chicken and ground turkey to at least 165°F.

- Cook beef, veal, and lamb roasts and steaks to an internal temperature of at least 145°F plus a 3-minute rest period, which leaves meat slightly pink in the center. Pork chops, roasts, and ribs should be cooked to at least 145°F plus a 3-minute rest period.

- Cook whole poultry to at least 165°F.

- Cook fish until it is opaque and flakes easily with a fork.

- Cook eggs until both the yolk and white are firm (especially for eggs that are pooled together before cooking; generally when making scrambled eggs). Since the *Salmonella* Enteritidis (bacterial serotype most frequently associated with foodborne illness cases involving shell eggs) contamination rate for shell eggs is very low (estimated at less than one egg in 20,000), the food safety risk associated with consuming over-easy eggs should be low, but it is never zero. Cook egg dishes to 160°F. Packaged liquid whole egg, egg yolks, and whites that have been pasteurized are safe when cooked to lower final temperatures.

- Reheat sauces, marinades, soups, and gravy to a rolling boil. Heat other leftovers and casseroles thoroughly to at least 165°F.

- Cook seafood to 145°F for finned fish; shellfish until they are pearly and opaque; clams, oysters, and mussels until the shell opens during cooking; and scallops until they are milky white or opaque and firm.

6. Serve food safely.
   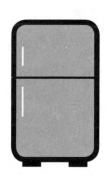
   - Serve food on clean dishes and utensils, not those used in food preparation.

   - Never leave perishable food out of the refrigerator for more than 2 hours. Depending on the outside temperature, if food is left out at a picnic or in a hot car, it may remain safe for only 30 minutes.

7. Handle leftovers properly.
   - Divide large amounts of leftovers into small, shallow containers for quick cooling in the refrigerator.

   - Remove stuffing from meats and poultry, and refrigerate it in a separate container. It is recommended that stuffing be cooked separately rather than inside of poultry.

   - Don't eat cooked or perishable foods that have been kept in the refrigerator for too long (no more than 2 or 3 days).

Never taste food that looks or smells strange to determine if you can still use it.

■ When in doubt, throw it out.

8. If you think you are sick from foodborne bacteria, let someone know.

Hey guys, that quiche might've made me sick.

■ If you are concerned or have questions about your health, consult your physician.

■ Any instance of diarrhea, vomiting, abdominal pain, or headache lasting longer than two days should be reported to a physician.

■ Most foodborne microorganisms take from approximately 1 to 3 days to cause symptoms. However, there are exceptions. *Staphylococcus* species can cause food intoxication within 1 to 6 hours of ingestion. Conversely, *Listeria monocytogenes* can cause listeriosis within 2 to 21 days. Disease onset depends on such factors as the general health status of the person, the dose (population) of bacteria in the food and portion of food consumed, and the virulence (infectivity) of the organism. If you see or talk to a physician, be prepared to give an account of all foods you have consumed over the past week or more.

SOME FINAL THINGS TO CONSIDER

If you have read this book from cover to cover, you must either have a lot of free time on your hands or be very interested in bacteria and how they are related to food safety. In any case, we hope you have found our book informative as well as entertaining. These topics may not be on the high-priority check-list for the food safety police, but perhaps we have raised your awareness that everyday practices like playing beer pong, double-dipping, or sharing popcorn at a movie theater do come with some potential safety risks. So take this knowledge and go forth. Be safe and healthy. Enjoy food and drink. And have fun!

Acknowledgments

FROM PAUL:

It is a unique opportunity to acknowledge many people that have helped me achieve something as noteworthy as this book. This list is too long to fit here, but be assured that I am forever grateful they have been part of my life. Most important to me is to dedicate this work to my wife, Rose, who is the love of my life (Halley's Comet) and my daughter, Elizabeth, who I unconditionally love (punks). Next on my dedication parade are my parents, Pauline and Sheldon, who I miss but am sure are watching, and my brother, Shelly, and nephew, Jason. Of course, Brian, my coauthor, who became my major professor nearly four decades ago somewhat by chance but subsequently became a close friend, going on many trips and runs with me; thanks for letting me twist your arm to write this book. Inyee Han has been a special colleague and has my everlasting gratitude for her dedicated work and friendship over the past quarter century. Jim,

239

Johnny, Julie, and many others at Clemson University have been good friends and colleagues. Thanks to the NC State Road Scholars and especially Tom and Steve. Morgan has been a great lifelong friend whose advice was integral in publishing this book. Farley, Quynh, and those at Norton have been fantastic in getting the book to the finish line. Brian C., your graphics are great. Finally, thanks to students who worked in my lab and contributed to research publications over the past 20+ years.

FROM BRIAN:

Thank you to the many people who have been so influential in my life and have contributed either directly or indirectly in helping me author this book. Foremost, I dedicate this book to my wife, Yvonna (Bonnie), who has not only loved me unconditionally but has been a constant companion, encourager, and supporter throughout our marriage. As a microbiologist herself, she has provided a critical eye in reviewing and improving the overall quality of this manuscript. I would not be where I am today without my parents, who nurtured me through those most impressionable years and instilled in me a strong work ethic and desire to succeed. Without question, the many undergraduate and graduate students that I have taught and mentored through the years have been foundational in my development as a teacher, scientist, and most importantly as a human being. It goes without question that without my close friend and coauthor (Dr. Dawson), this book would have been a distant dream. What started as a means for inspiring undergraduate students to participate in scientific inquiry through laboratory research has led to a flood of public interest in its content and ultimately to the writing of this book. Finally, I give my heartfelt love and thanks to my Lord and Savior, Jesus Christ, to whom I owe everything.

Notes

- - - - - - - - - - -

INTRODUCTION: A DIVE INTO THE MYSTERIOUS MICROBIAL WORLD

1. Whitman, W. B., Coleman, D. C., & Wiebe, W. J. (1998). Prokaryotes: The unseen majority. *Proceedings of the National Academy of Sciences, USA, 95*(12), 6578–6583; http://www.cen.ulaval.ca/merge/pdf/Whitman1998.pdf

2. Lougheed, K. (2012, August). There are fewer microbes out there than you think. *Nature.* Retrieved from https://www.nature.com/news/there-are-fewer -microbes-out-there-than-you-think-1.11275

3. Haub, C. (2011). How many people have ever lived on earth? *Population Reference Bureau.* Retrieved from http://www.prb.org/Publications/Articles/2002/ HowManyPeopleHaveEverLivedonEarth.aspx

4. Rappe, M. S., & Giovannoni, S. J. (2003). The uncultured microbial majority. *Annual Review of Microbiology, 57,* 369–394.

5. Madigan, M. T., & Martinko, J. M. (2006). *Brock biology of microorganisms.* Upper Saddle River, NJ: Pearson Prentice Hall (p. 29).

6. Ibid. (p. 9).

7. Museum of Microscopy. (n.d.). Leeuwenhoek microscope. Retrieved from http://micro.magnet.fsu.edu/primer/museum/leeuwenhoek.html

8. Porter, J. R. (1976). Antony van Leeuwenhoek: Tercentenary. *Bacteriological Reviews, American Society of Microbiology, 40*(2), 260–269.

9. Bacteria. (n.d.). *Wikipedia*. Retrieved from https://www.wikipedia.org/bacteria

10. Ibid.

11. See note 5, p. 64.

12. Mandal, A. (2013). Human diseases caused by viruses. *News Medical Life Sciences*. Retrieved from https://www.news-medical.net/health/Human -Diseases-Caused-by-Viruses.aspx

13. Centers for Disease Control and Prevention (CDC). (n.d.). Foodborne illnesses and germs. Retrieved from https://www.cdc.gov/foodsafety/foodborne -germs.html

14. Breitbart, M., & Rohwer, F. (2005). Here a virus, there a virus, everywhere the same virus? *Trends in Microbiology, 13*(6), 278–284.

15. Edwards, R. A., & Rohwer, F. (2005). Viral metagenomics. *Nature Reviews Micro- biology, 3*(6), 504–510.

16. See note 5, p. 231.

17. Jingnan, L., Brihman, C. J., Gai, C. S., & Sinskey, A. J. (2012). Studies on the pro- duction of branched-chain alcohols in engineered *Ralstonia eutropha*. *Applied Microbiology and Biotechnology, 96*(1), 283–297.

18. Encyclopaedia Britannica. (n.d.). Nitrogen-fixing bacteria. Retrieved from https://www.britannica.com/science/nitrogen-fixing-bacteria

19. See note 5, pp. 102–105.

20. See note 5, pp. 157–158.

21. See note 5, pp. 150–152.

22. Eagon, R. G. (1962). *Pseudomonas natriegens*, a marine bacterium with a gen- eration time of less than 10 minutes. *Journal of Bacteriology, 83*(4), 736–737.

23. See note 5, pp. 136–137.

24. Centers for Disease Control and Prevention (CDC). (2015, May). Preliminary incidence and trends of infection with pathogens transmitted commonly through food. *Mortality and Morbidity Weekly Report, 64*(18), 495–499. Retrieved from https://www.cdc.gov/mmwr/pdf/wk/mm6418.pdf

25. CDC. (n.d.). Foodborne illnesses and germs. Retrieved from https://www
.cdc.gov/foodsafety/foodborne-germs.html
26. CDC. (n.d.). Foodborne outbreaks. Retrieved from http://www.cdc.gov/
foodsafety/outbreaks/index.html
27. CDC. (n.d.). Burden of foodborne illness: Findings. Retrieved from http://
www.cdc.gov/foodborneburden/2011-foodborne-estimates.html
28. Bottemiller, H. (2012, January). Annual foodborne illnesses cost $77 bil-
lion, study finds. *Food Safety Magazine*. Retrieved from http://www.food
safetynews.com/2012/01/foodborne-illness-costs-77-billion-annually-study
-finds/#.VqZGgCorKUk
29. Gillespie, P., & Alesci, C. (2016, February). Chipotle profits tank after *E. coli*
scare. *CNN Money*. Retrieved from http://money.cnn.com/2016/02/02/
investing/chipotle-earnings-e-coli/index.html
30. Maberry, T. (2017, February). A look back at 2016 food recalls. *Food Safety
Magazine*. Retrieved from http://www.foodsafetymagazine.com/A Look Back
at 2016 Food Recalls.html
31. Tyco Integrated Security. (2012, October). Recall: The food industry's
biggest threat to profitability. *Food Safety Magazine*. Retrieved from
https://www.foodsafetymagazine.com/signature-series/recall-the-food
-industrys-biggest-threat-to-profitability/
32. Hussain, M. A., & Dawson, C. O. (2013). Economic impact of food safety outbreaks on food businesses. *Foods, 2*, 585–589.
33. Tyco Integrated Security. (2012, October). "Recall: The food industry's biggest threat to profitability," *Food Safety Magazine*. https://www
.foodsafetymagazine.com/signature-series/recall-the-food-industrys
-biggest-threat-to-profitability/
34. Foodborne viruses. (2008, April). *News Medical Life Sciences*. Retrieved from
http://www.news-medical.net/news/2008/04/09/37149.aspx
35. Ibid.

PART 1: SURFACES

1. Costerton, J. W., Geesey, G. G., & Cheng, K. J. (1978). How bacteria stick. *Scientific American, 238*, 86–95.

2. Alhede, M., Jensen, P. O., Givskov, M., & Bjarnsholt, T. (2018). Biofilm of medical importance (Biotechnology Vol. XII). In *Encyclopedia of life support systems (EOLSS)*. United Nations Educational Scientific and Cultural Organization (UNESCO). http://www.eolss.net/Sample-Chapters/C17/E6-58-11 -12.pdf

3. Davies, D. (2003). Understanding biofilm resistance to antibacterial agents. *Nature, 2*, 114–122.

4. Costerton, J. W., Stewart, P. S., & Greenberg, E. P. (1999). Bacterial biofilms: A common cause of persistent infections. *Science, 284*, 1318–1322.

5. Reynolds, K. A., Watt, P. A., Boone, S. A., & Gerba, C. P. (2005). Occurrence of bacteria and biochemical markers on public surfaces. *International Journal of Environmental Health Research, 15*(3), 225–243.

6. Chawla, K., Mukhopadhayay, C., Gurung, B., Bhate, P., & Bairy, I. (2009). Bacterial "cell" phones: Do cell phones carry potential pathogens? *Online Journal of Health and Allied Sciences, 8*(1), 8. Retrieved from http:// cogprints.org/6566/1/2009-1-8.pdf

7. Dunkin, M. A. (2009). 6 surprisingly dirty places in your home. *WebMD*. Retrieved from http://www.webmd.com/women/home-health-and-safety-9/ places-germs-hide?page=4

CHAPTER 1: THE FIVE-SECOND RULE

1. *Wikipedia*. (2005). Retrieved from https://en.wikipedia.org/wiki/Five-second _rule

2. Strout, S. (2001). Does the five second rule exist? *Fact Index*. Retrieved from http://www.fact-index.com/f/fi/five_second_rule.html; Naish, J. (2016). Why eating food off the floor can be good for you: Scientists have warned anything that touches the floor is at risk of *E. coli* or salmonella—but are they right? Retrieved from http://www.dailymail.co.uk/news/article-3518327/Why-eating -food-floor-good-Scientists-warned-touches-floor-risk-E-colli-salmonella -right.html

3. Leach, B. (2006, December). Does the five-second rule really work? Retrieved from http://scienceline.org/2006/12/ask-leach-fiveseconds/

4. Dawson, P. L., & Schanzle, J. (2015, September). Explainer: Is it really OK to eat food that's fallen on the floor? *The Conversation* (online news website).

Retrieved from https://theconversation.com/explainer-is-it-really-ok-to-eat -food-thats-fallen-on-the-floor-45541

5. Julia flubs her flip. (2011, September). *YouTube*. Retrieved from https:// www.youtube.com/watch?v=iR64GGyEv_o

6. Jacobs, L. (2009, August). Our lady of the kitchen. Retrieved from http:// www.vanityfair.com/culture/2009/08/julia-child200908

7. Dawson, P., Han, I., Cox, M., Black, C., & Simmons, L. (2006). Residence time and food contact time effects on transfer of *Salmonella typhimurium* from tile, wood and carpet: Testing the five-second rule. *Journal of Applied Microbiology, 102*(4), 945–953; Miranda, R., & Schaffner, D. (2016, September). Longer contact times increase cross-contamination of *Enterobacter aerogenes* from surfaces to food. *Applied Environmental Microbiology, 82*(21), 6490–6496. Retrieved from http://aem.asm.org/content/early/2016/08/15/AEM.01838 -16.full.pdf

8. College of Agricultural, Consumer and Environmental Sciences (ACES). (2003, September). If you drop it, should you eat it? Scientists weigh in on the 5-second rule. Retrieved from http://news.aces.illinois.edu/news/if-you-drop -it-should-you-eat-it-scientists-weigh-5-second-rule

9. The five-second rule. (2005). *Annotated MythBusters*. Retrieved from http:// kwc.org/mythbusters/2005/10/mythbusters_chinese_invasion_a.html

10. See note 7.

11. Connecticut College. (2018). Chemistry professor awarded $300K to continue bioluminescence research. Retrieved from https://www.conncoll.edu/news/ search-results/?q=https%3A%2F%2Fwww.conncoll.edu%2Fnews%2Fnews -archive%2F2018%2Fbranchini-grant

12. Aston University. (2014, March). Researchers prove the five second rule is real. News release; retrieved from http://www.aston.ac.uk/about/news/ releases/2014/march/five-second-food-rule-does-exist/

13. Is the "5 second rule" legit? (n.d.). *The Quick and the Curious*. The Science Channel, retrieved from http://www.sciencechannel.com/tv-shows/ the-quick-and-the-curious/the-quick-and-the-curious-videos/is-the-5 -second-rule-legit/

14. See note 7.

15. Statistical Analysis System (SAS). (2015). SAS OnDemand for Academics. https://www.sas.com/en_us/software/on-demand-for-academics.html

16. Paustian, T. (2013, May). Endospores are very resistant structures. In *Through the microscope: A microbiology textbook*. Retrieved from http://www.microbiologytext.com/5th_ed/book/displayarticle/aid/69

17. Humphrey, T. J., Martin, K. W., & Whitehead, A. (1994). Contamination of hands and work surfaces with *Salmonella enteriditis* PT4 during the preparation of egg dishes. *Epidemiology and Infection, 113*, 403–409; Scott, E., & Bloomfield, S. (1990). The survival and transfer of microbial contamination via cloth, hands, and utensils. *Journal of Applied Bacteriology, 68*, 271–278.

18. Kusumaningrum, H. D., Riboldi, G., Hazelberger, W. C., & Beumer, R. R. (2003). Survival of foodborne pathogens on stainless steel surfaces and cross-contamination to foods. *International Journal of Food Microbiology, 85*, 227–236.

19. Moore, C. M., Sheldon, B. W., & Jaykus, L-A. (2003). Transfer of *Salmonella* and *Campylobacter* from stainless steel to romaine lettuce. *Journal of Food Protection, 66*, 2231–2236.

20. Chen, Y., Jackson, M., Chen, F., & Schaffner, D. (2001). Quantification and variability analysis of bacterial cross-contamination rates in common food service tasks. *Journal of Food Protection, 64*, 72–80.

21. See note 19.

22. Browdie, B. (2012, April). Gross! America's dogs poop 10 million tons a year. *New York Daily News*. Retrieved from http://www.nydailynews.com/news/national/america-dogs-poop-10-million-tons-year-potential-health-hazard-waste-firm-finds-article-1.1065362; Piddle Place. (2016, May). How much should my dog urinate? [Blog post by Steven Rowley]. Retrieved from https://www.piddleplace.com.hk/blogs/news/124312771-how-much-should-my-dog-urinate

CHAPTER 2: BEER PONG: DON'T HATE THE GAME

1. Berner, L. (2004, November). On language, Princeton style: The history of "Beirut." *Daily Princetonian*. Retrieved from https://archive.li/pLiJz

2. Statistical Analysis System (SAS). (2015). SAS OnDemand for Academics. https://www.sas.com/en_us/software/on-demand-for-academics.html

3. Hui, Y. H., & Sherkat, F. (2000). *Handbook of food science, technology and engineering*. New York, NY: Taylor and Francis.

4. Allowed, P. B., Jenkins, T., Paulus, L. N., Johnson, L., & Hedberg, C. W. (2004).

Hand washing compliance among retail food establishment workers in Minnesota. *Journal of Food Protection, 67*(12), 2825–2828; Daniels, N. A., Mackinnon, L., Rowe, S., Bean, N. H., Griffin, P. M., & Mead, P. S. (2002). Foodborne disease outbreaks in United States schools. *Pediatric Infectious Disease Journal, 21*(7), 623–628; Fry, A. M., Braden, C. R., Griffin, P. M., & Hughes, J. M. (2005). Foodborne diseases. In G. L. Mandell, J. E. Bennett, & R. Dolin (Eds.), *Principles and Practice of Infectious Diseases.* 6th ed. (pp. 1286–1297). New York, NY: Elsevier; Shojaei, H. J., Shooshtaripoor, J., & Amiri, M. (2006). Efficacy of simple hand washing in reduction of microbial hand contamination of Iranian food handlers. *Food Research International, 39,* 525–529.

5. See note 2.
6. Morton, H. (1950). The relationship of concentration and germicidal efficiency of ethyl alcohol. *Annals of the New York Academy of Sciences, 53*(1), 191–196.
7. Perez-Rodriguez, F., Valero, A., Carrasco, E., Garcia, R. M., & Zurera, G. (2008). Understanding and modelling bacterial transfer to foods: A review. *Trends in Food Science and Technology, 19*(3), 131–144.

CHAPTER 3: ARE YOU READY TO ORDER?

1. Aycicek, H., Oguz, U., & Karci, K. (2006). Comparison of results of ATP bioluminescence and traditional hygiene swabbing methods for the determination of surface cleanliness at a hospital kitchen. *International Journal of Hygiene and Environmental Health, 209*(2), 203–206.
2. National Restaurant Association. (n.d.). Facts at a glance. Retrieved from http://www.restaurant.org/News-Research/Research/Facts-at-a-Glance
3. Statista, The Statistics Portal. (n.d.). Eating out behavior in the U.S.—statistics & facts. Retrieved from https://www.statista.com/topics/1957/eating-out-behavior-in-the-us
4. Centers for Disease Control and Prevention (CDC). (2016). Foodborne Outbreak Online Database (FOOD Tool)—United States, 1998–2016. Retrieved from https://wwwn.cdc.gov/foodborneoutbreaks/
5. Redmond, E. C., & Griffith, C. J. (2003). Consumer food handling in the home: A review of food safety studies. *Journal of Food Protection, 66*(1), 130–161.
6. Tirado, C., & Schmidt, K. (2001). *The WHO Surveillance Programme for Control of Foodborne Infections and Intoxications: Preliminary Results and Trends Across*

Greater Europe. Journal of Infection, 43, 80–84. Retrieved from https://ac.els
-cdn.com/S0163445301908618/1-s2.0-S0163445301908618-main.pdf?_
tid=8d4b97a1-197f-4d88-837a-e04f73405a4d&acdnat=1520263856_d9e
86ab20e63865faa5d26856711a403; Olsen, S. J., MacKinnon, L., Goulding,
J., Bean, N. H., & Slutsker, L. (2000). Surveillance for foodborne-disease outbreaks: United States, 1993–1997. *Morbidity and Mortality Weekly Report*,
CDC Surveillance Summary; *49*(SS-1), 1–62.

7. Green, L., Selman, C., Banerjee, A., Marcus, R., Medus, C., Angulo, F. J., &
Buchanan, S. (2005). Food service workers' self-reported food preparation
practices: An EHS-Net study. *International Journal of Hygiene and Environmental Health, 208*(1), 27–35.

8. Simon, P. A., Leslie, P., Run, G., Jin, G. Z., Reporter, R., Aguirre, A., & Fielding, J. E. (2005). Impact of restaurant hygiene grade cards on foodborne-disease
hospitalizations in Los Angeles County. *Journal of Environmental Health, 67*(7),
32–36.

9. Taku, A., Gulati, B. R., Allwood, P. B., Palazzi, K., Hedberg, C. W., &
Goyal, S. M. (2002). Concentration and detection of caliciviruses from food
contact surfaces. *Journal of Food Protection, 65*(6), 999–1004.

10. Wheeler, C., Vogt, T. M., Armstrong, G. L., Vaughan, G., Weltman, A., Nainan,
O. V., . . . Bell, B. P. (2005). An outbreak of hepatitis A associated with green
onions. *New England Journal of Medicine, 353*, 890–897. Retrieved from http://
www.nejm.org/doi/full/10.1056/NEJMoa050855

11. Henson, S., Majowicz, S., Masakure, O., Sockett, P., Jones, A., Hart, R., &
Knowles, L. (2006). Consumer assessment of the safety of restaurants: The
role of inspection notices and other information cues. *Journal of Food Safety*,
26(4), 275–301.

12. Cruz, M. A., Katz, D. J., & Suarez, J. A. (2001). An assessment of the ability of
routine restaurant inspections to predict food-borne outbreaks in Miami–Dade
County, Florida. *American Journal of Public Health, 91*(5), 821.

13. Moore, G., & Griffith, C. (2002). A comparison of traditional and recently developed methods for monitoring surface hygiene within the food industry: An industry trial. *International Journal of Environmental Health Research, 12*(4), 317–329.

14. Clayton, D. A., & Griffith, C. J. (2004). Observation of food safety practices
in catering using notational analysis. *British Food Journal, 106*(3), 211–227;
Marler Clark. (2006). Retrieved from https://marlerclark.com/

15. Knight, A. J., Worosz, M. R., & Todd, E. C. D. (2007). Serving food safety: Consumer perceptions of food safety at restaurants. *International Journal of Contemporary Hospitality Management, 19*(6), 476–484.

16. Buchholz, U., Run, G., Kool, J. L., Fielding, J., & Mascola, L. (2002). A risk-based restaurant inspection system in Los Angeles County. *Journal of Food Protection, 65*(2), 367–372.

17. Teixeira, P., Silva, S. C., Araújo, F., Azeredo, J., & Oliveira, R. (2007). Bacterial adhesion to food contacting surfaces. In A. Mendez-Vilas (Ed.), *Communicating Current Research and Educational Topics and Trends in Applied Microbiology* (pp. 13–20). Open Access Formatex. Retrieved from http://www.formatex.org/microbio/pdf/Pages13-20.pdf

18. Sirsat, A., Choi, J. K., & Neal, J. (2013). Persistence of *Salmonella* and *E. coli* on the surface of restaurant menus. *Journal of Environmental Health, 75*(7), 8–16.

19. Choi, J. K., Almanza, B., Nelson, D. Neal, J., & Sirsat, S. (2014). A strategic cleaning assessment program: Menu cleanliness at restaurants. *Journal of Environmental Health, 76*(10), 18–24.

20. Clemons, J. A. (2010). Novel approaches for the efficient sampling and detection of *Listeria monocytogenes* and *Brochothrix thermosphacta* on food contact surfaces (unpublished master's thesis). Knoxville: University of Tennessee. Retrieved from http://trace.tennessee.edu/utk_gradthes/784/

21. Statistical Analysis System (SAS). (2015). SAS OnDemand for Academics. Retrieved from https://www.sas.com/en_us/software/on-demand-for-academics.html

22. Ibid.

23. Ibid.

24. Milling, A., Kehr, R., Wulf, A., & Smalla, K. (2005). Survival of bacteria on wood and plastic particles: Dependence on wood species and environmental conditions. *Holzforschung, 59*(1), 72-81; Torres, A. G., Jeter, C., Langley, W., & Matthysse, A. G. (2005). Differential binding of *Escherichia coli* O157: H7 to alfalfa, human epithelial cells, and plastic is mediated by a variety of surface structures. *Applied and Environmental Microbiology, 71*(12), 8008–8015.

25. U.S. Food and Drug Administration (FDA). (n.d.). Bad bug book. Retrieved from https://www.fda.gov/Food/FoodborneIllnessContaminants/CausesOfIllnessBadBugBook/

26. Baron, F., Gautier, M., & Loir, Y. (2003). *Staphylococcus aureus* and food poisoning. *Genetics and Molecular Research, 2*, 63–76.

27. Neely, A. N., & Maley, M. P. (2000). Survival of enterococci and staphylococci on hospital fabrics and plastic. *Journal of Clinical Microbiology, 38*(2), 724–726.

28. Schreck, S. (2009, May). Bacteria on the menu? *Food Poison Journal.* Retrieved from http://www.foodpoisonjournal.com/food-poisoning-watch/bacteria-on -the-menu/#.WVvZ1YWcHIW

29. Cleaning & Maintenance Management (CMM). (2014). Beware of germs on the menu. Retrieved from http://www.cmmonline.com/articles/233211 -beware-of-germs-on-the-menu

PART 2: AIR & WATER

1. Wan, G. H., Wu, C. L., Chen, Y. F., Huang, S. H., Wang, Y. L., & Chen, C. W. (2014). Particle size concentration distribution and influences on exhaled breath particles in mechanically ventilated patients. *PLoS ONE, 9*(1), e87088.

2. Chao, C. Y. H., Wan, M. P., Morawska, L., Johnson, G. R., Ristovski, Z. D., Hargreaves, K., . . . & Katoshevski, D. (2009). Characterization of expiration air jets and droplet size distributions immediately at the mouth opening. *Journal of Aerosol Science, 40*(2), 122–133.

3. Madigan, M. T., Martinko, J. M., Dunlap, P. V., & Clark, D. P. (2009). *Brock biology of microorganisms.* 12th ed. San Francisco, CA: Pearson–Benjamin Cummings (p. 1061).

4. Weber, T. P., & Stilianakis, N. I. (2008). Inactivation of influenza A viruses in the environment and modes of transmission: A critical review. *Journal of Infection, 57*, 361–373; Wein, L. M., & Atkinson, M. P. (2009). Assessing infection control measures for pandemic influenza. *Risk Analysis, 29*, 949–962.

5. Fabian, P., McDevitt, J. J., DeHaan, W.H., Fung, R.O.P., Cowling, B.J., Chan, . . . Milton, D. K. (2008). Influenza virus in human exhaled breath: An observational study. *PLoS ONE, 3*, e2691; Lindsley, W. G., Blachere, F. M., Thewlis, R. E., Vishnu, A., Davis, K. A., Cao, G., & Palmer, J. E. (2010). Measurements of airborne influenza virus in aerosol particles from human coughs. *PLoS ONE, 5*, e15100.

6. Fennelly, K. P., Martyny, J. W., Fulton, K. E., Orme, I. M., Cave, D. M., & Heifets, L. B. (2004). Cough-generated aerosols of *Mycobacterium tuberculosis*:

A new method to study infectiousness. *American Journal of Respiratory and Critical Care Medicine, 169*(5), 604–609.

7. Vollaard, A. M., Soegianto, A., van Asten, H., Suwandhi, W., Visser, L., Surjadi, C., & van Dissel, J. (2004). Risk factors for typhoid and paratyphoid fever in Jakarta, Indonesia. *Journal of the American Medical Association, 291,* 2607–2615.

8. Kerr, K. G., Seale, K., Walbran, S., Rajgopal, A., & Bentham, D. (2003). Prevalence of *Escherichia coli* O157:H7 on the hands of food-workers. *British Food Journal, 105,* 678–681; Lane, C. (2001). Bacteria can be a real handful. *Food Processing, 70,* 21; Michaels, B. (2002). Handwashing: An effective tool in the food safety arsenal. *Food Quality, 9,* 45, 46, 49–53; Michaels, B., Gangar, V., Schultz, A., Arenas, M., Curiale, M., Ayers, T., & Paulson, D. (2002). Water temperature as a factor in handwashing efficacy. *Food Service Technology, 2,* 139–149; Paulson, D. S. (2000). Handwashing, gloving, and disease transmission by the food preparer. *Dairy, Food and Environmental Sanitation, 20,* 838–845; Paulson, D. S., Riccardi, C., Beausoleil, C. M., Fendler, E. J., Dolan, M. J., Dunkerton, L. V., & Williams, R. A. (1999). Efficacy evaluation of four hand cleansing regimens for food handlers. *Dairy, Food and Environmental Sanitation, 19,* 680–684; Taylor, A. K. (2000). Food protection: New developments in handwashing. *Dairy, Food and Environmental Sanitation, 20,* 114–119.

9. Bennett, T. (2003). Please don't bug me. *Food Processing, 72,* 15–16; Featherstone, S. (2003). Food hygiene—not for sissies. *Food Review, 30,* 47, 49; Lane, C. (2001). Bacteria can be a real handful. *Food Processing, 70,* 21.

10. Califano, A. N., deAntoni, G. L., Giannuzzi, L., & Mascheroni, R. H. (2000). Prevalence of unsafe practices during home preparation of food in Argentina. *Dairy, Food and Environmental Sanitation, 20,* 934–943; Clayton, D. A., Griffith, C. J., & Price, P. (2003). An investigation of the factors underlying consumers' implementation of specific food safety practices. *British Food Journal, 105,* 434–453; Kassa, H., Harrington, B., Bisesi, M., & Khuder, S. (2001). Comparisons of microbiological evaluation of selected kitchen areas with visual inspections for preventing potential risk of foodborne outbreaks in food service operations. *Journal of Food Protection, 64,* 509–513; Kohl, K. S., Rietberg, K., Wilson, S., & Farley, T. A. (2002). Relationship between home food-handling practices and sporadic salmonellosis in adults in Louisiana, United States. *Epi-*

demiology and Infection, 129, 267–276; Schaffner, D. W. (2003). Challenges in cross-contamination modeling in home and food service settings. *Food Australia*, 55, 583–586; Toshima, Y., Ojima, M., Yamada, H., Mori, H., Tonomura, M., Hioki, Y., & Koya, E. (2001). Observation of everyday handwashing behavior of Japanese, and effects of antibacterial soap. *International Journal of Food Microbiology*, 68, 83–91.

11. Worsfold, D. (2003). Food safety at shows and fairs. *Nutrition and Food Science*, 33, 159–164.

12. Albrecht, J. A., Sumner, S., & Henneman, A. (1992). Food safety in child care facilities. *Dairy, Food and Environmental Sanitation*, 12, 740–743.

13. Montville, R., Chen, Y., & Schaffner, D. W. (2002). Risk assessment of hand washing efficacy using literature and experimental data. *International Journal of Food Microbiology*, 73(2), 305–313.

14. Fendler, E. J., Dolan, M. J., & Williams, R. A. (1998). Handwashing and gloving for food protection. I. Examination of the evidence. *Dairy, Food and Environmental Sanitation*, 18, 814–813; Fendler, E. J., Dolan, M. J., Williams, R. A., & Paulson, D. S. (1998). Handwashing and gloving for food protection: II. Effectiveness. *Dairy, Food and Environmental Sanitation*, 18, 824–829; Guzewich, J. J. (1995). Bare hand contact with food, why isn't handwashing good enough? *Conference Proceeding of the International Association of Milk, Food and Environmental Sanitarians*, 58 (Supplement), 60; Montville, R., Yuhuam, C., & Schaffner, D. W. (2001). Glove barriers to bacterial cross-contamination between hands and food. *Journal of Food Protection*, 64, 845–849; Paulson, D. S. (1994). A comparative evaluation of different hand cleansers. *Dairy, Food and Environmental Sanitation*, 14, 524–528.

15. Michaels, B., Gangar, V., Chia-Min, M., & Doyle, M. (2003). Use limitations of alcoholic sanitizer as part of a food service hand hygiene program. *Food Service Technology*, 3, 71–80; Miller, M. L., James-Davis, L. A., & Milanesi, L. E. (1994). A field study evaluating the effectiveness of different hand soaps and sanitizers. *Dairy, Food and Environmental Sanitation*, 14, 155–160; Morita, S., Maeda, T., Taniguchi, R., Nakamura, M., Tachibana, M., Miyazaki, H., . . . Kumagai, S. (1999). Study for effectiveness of washing hands with various soaps and disinfectants. *Japanese Journal of Food Microbiology*, 16, 65–70; Paulson, D. S. (1994). A comparative evaluation of different hand cleansers. *Dairy, Food and Environmental Sanitation*, 14, 524–528; Toshima, Y., Ojima, M.,

Yamada, H., & Koya, E. (2001). Observation of everyday handwashing behavior of Japanese, and effects of antibacterial soap. *International Journal of Food Microbiology, 68*, 83–91.

16. Michaels, B., Gangar, V., Schultz, A., Arenas, M., Curiale, M., Ayers, T., & Paulson, D. 2001. Handwashing water temperature effects on the reduction of resident and transient (*Serratia marcescens*) flora when using bland soap. *Dairy, Food and Environmental Sanitation, 21*, 9997–1007; Jensen, D., Macinga, D., Shumaker, D., Belino, R., Arbogast, J., & Schaffner, D. 2017. Quantifying the effects of water temperature, soap volume, lather time, and antimicrobial soap as variables in the removal of *Escherichia coli* ATCC 11229 from hands. *Journal of Food Protection, 80*, 1022–1031.

CHAPTER 4: BLOWING OUT BIRTHDAY CANDLES, OR SPRAYING GERMS ON CAKE?

1. Frey, A. (1753). *A true and authentic account of Andrew Frey.* Retrieved from https://books.google.com/books?id=VIoUAAAAQAAJ&pg=PA15&hl=en#v=o nepage&q&f=false

2. Fabian, P., McDevitt, J. J., DeHaan, W. H., Fung, R. O. P., Cowling, B. J., Chan, K. H. . . . Milton, D. K. (2008). Influenza virus in human exhaled breath: An observational study. *PLoS ONE, 3*, e2691; Stelzer-Braid, S., Oliver, B. G., Blazey, A. J., Argent, E., Newsome, T. P., Rawlinson, W. D., & Tovey, E. R. (2009). Exhalation of respiratory viruses by breathing, coughing, and talking. *Journal of Medical Virology, 81*, 1674–1679; Huynh, K. N., Oliver, B. G., Stelzer, S., Rawlinson, W. D., & Tovey, E. R. (2008). A new method for sampling and detection of exhaled respiratory virus aerosols. *Clinical Infection and Disease, 46*, 93–95.

3. Obeng, C. S. (2008). Personal cleanliness activities in preschool classrooms. *Early Childhood Education Journal, 36*, 93–99.

4. Xu, Z., Shen, F., Li, X., Wu, Y., Chen, Q., Jie, X., & Yao, M. (2012). Molecular and microscopic analysis of bacteria and viruses in exhaled breath collected using a simple impaction and condensing method. *PloS ONE, 7*(7), 1–8. Retrieved from http://www.plosone.org; Qian, J., Hospodsky, D., Yamamoto, N., Nazaroff, W. W., & Peccia, J. (2012). Size-resolved emission rates of airborne bacteria and fungi in an occupied classroom. *Indoor Air, 22*(4), 339–351.

5. Statistical Analysis System (SAS). (2015). SAS OnDemand for Academ-

ics. Retrieved from https://www.sas.com/en_us/software/on-demand-for
-academics.html

6. Flügge, C. (1899). Die verbreitung der phthise durch staubförmiges sputum und durch beim husten verspritzte tröpfchen. [The dissemination of phthisis by dusty sputum and by the droplets spilled on coughing]. *Zeitschrift fur Hygiene und Infektionskrankheiten, 30*(1), 107–124; Hutchison, R. F. (1901). Die verbreitung von keimen durch gewöhnliche luftströme. [The spread of germs by ordinary air currents]. *Zeitschrift fur Hygiene und Infektionskrankheiten, 36*(1), 223–253.

7. Duguid, J. P. (1946). The size and the duration of air-carriage of respiratory droplets and droplet nuclei. *Epidemiology and Infection, 44*(6), 471–479.

8. Douwes, J., Thorne, P., Pearce, N., & Heederik, D. (2003). Bioaerosol health effects and exposure assessment: Progress and prospects. *Annals of Occupational Hygiene, 3*, 187–200.

9. Houk, V. N. (1980). Spread of tuberculosis via recirculated air in a naval vessel: The Byrd study. *Annals of the New York Academy of Sciences, 353*, 10–24; Loosli, C., Hertweck, M., & Hockwald, R. (1970). Airborne influenza PR8-A virus infections in actively immunized mice. *Archives of Environmental Health, 21*, 332–346.

10. Yu, I. T., Li, Y., Wong, T. W., Tam, W., Phil, M., Chan, A. T., . . . Ho, T. (2004). Evidence of airborne transmission of the severe acute respiratory syndrome virus. *New England Journal of Medicine, 351*, 1731–1739; Brankston, G., Gitterman, L., Hirji, Z., Lemieux, C., & Gardam, M. (2007). Transmission of influenza A in human beings. *Lancet Infectious Diseases, 7*, 257–265.

CHAPTER 5: KEEP YOUR DIRTY HANDS TO YOURSELF

1. Borchgrevinik, C., & Henion, A. (2013, June). Eww! Only 5 percent wash hands correctly. *MSU Today*. Retrieved from http://msutoday.msu.edu/news/2013/eww-only-5-percent-wash-hands-correctly/

2. Boshell, P. (2016). Can the "yuck factor" motivate people to wash their hands? [Hand hygiene, infection prevention, and food safety blog]. Retrieved from http://info.debgroup.com/blog/can-the-yuck-factor-motivate-people-to-wash-their-hands?utm_campaign=blog&utm_content=35190984&utm_medium=social&utm_source=linkedin

3. U.S. Department of Labor, Bureau of Labor Statistics. (2014). *Occupational Outlook Handbook*, Food and Beverage Serving and Related Workers. Retrieved

from http://www.bls.gov/ooh/Food-Preparation-and-Serving/Food-and -beverage-serving-and-related-workers.htm

4. National Center for Public Policy and Higher Education. (2005, November). *Policy Alert*. Retrieved from http://www.highereducation.org/reports/pa_decline/ pa_decline.pdf

5. Hansen, T. B., & Knochel, S. (2003). Image analysis method for evaluation of specific and non-specific hand contamination. *Journal of Applied Microbiology, 94,* 483–494.

6. Marriot, N., & Gravari, B. (2006). *Principles of food sanitation.* 5th ed., Food Science Text Series. New York, NY: Springer (p. 83).

7. McElroy, D. M., & Cutter, C. N. (2004). Self-reported changes in food safety practices as a result of participation in a statewide food safety certification program. *Food Protection Trends, 24,* 150–161.

8. Doyle, M. P., Ruoff, K. L., Pierson, M., Soule, B., & Michaels, B. S. (2000). Reducing transmission of infectious agents in the home II. Control. *Dairy, Food and Environmental Sanitation, 20,* 418–425.

9. Toshima, Y., Ojima, M., Yamada, H., Mori, H., Tonomura, M., Hioki, Y., & Koya, E. (2001). Observation of everyday hand-washing behavior of Japanese and effects of antibacterial soap. *International Journal of Food Microbiology, 68,* 83–91.

10. Courtenay, M., Ramirez, L., Cox, B., Han, I., Jiang, X., & Dawson, P. (2005). Effect of various hand hygiene regimes on removal and/or destruction of *Escherichia coli* on hands. *Food Service Technology, 5,* 77–84.

11. Jiang, X., J. Morgan, & M. P. Doyle. (2002). Fate of *Escherichia coli* O157:H7 in manure-amended soil. *Applied and Environmental Microbiology, 68,* 2605–2609.

12. Statistical Analysis System (SAS). (2015). SAS OnDemand for Academics. Retrieved from https://www.sas.com/en_us/software/on-demand-for -academics.html

13. Michaels, B., Gangar, V., Schultz, A., Arenas, M., Curiale, M., Ayers, T., & Paulson, D. (2002). Water temperature as a factor in handwashing efficacy. *Food Service Technology, 2,* 139–149.

14. See note 12.

15. Michaels, B., Gangar, V., Chia-Min, M., & Doyle, M. (2003). Use limitations of alcoholic sanitizer as part of a food service hand hygiene program. *Food Service Technology, 3,* 71–80.

16. Kramer, A. P., Rudolph, P., Kampf, G., & Pittet, D. (2002). Limited efficacy of alcoholic based hand gels. *Lancet, 359,* 1489–1490.

17. Charbonneau, D. C., Ponte, J. M., & Kochanowski, B. A. (2000). A method for assessing the efficacy of hand sanitizers: Use of real soil encountered in the food service industry. *Journal of Food Protection, 63,* 495–501.

18. Sandora, T. J., Taveras, E. M., Shih, M-C., Resnick, E. A., Lee, G. M., & Ross-Degnan, D. (2004, September). Hand sanitizer reduces illness transmission in the home [Abstract 106]. In *Abstracts of the 42nd annual meeting of the Infectious Disease Society of America; Boston, Massachusetts; 2004 Sept 30–Oct 3.* Alexandria, VA: Infectious Disease Society of America; Hammond, B., Ali, Y., Fendler, E., Dolan, M., & Donovan, S. (2000). Effect of hand sanitizer use on elementary school absenteeism. *Journal of Infection Control, 28,* 340–346; White, C., Kolble, R., Carlson, R., Lipson, N., Dolan, M., & Ali, Y. (2003). The effect of hand hygiene on illness rate among students in university residence halls. *American Journal of Infection Control, 31,* 364–370.

19. Reynolds, S. A., Levy, F., & Walker, E. S. (2006). Hand sanitizer alert. *Emerging Infectious Diseases, 12*(3): 527–529.

20. Moore, C. M., Sheldon, B. W., & Jaykus, L. E. (2003). Transfer of *Salmonella* and *Campylobacter* from stainless steel to romaine lettuce. *Journal of Food Protection, 12,* 2231–2236.

21. Code of Federal Regulations. 21 CFR 310. 2016 Consumer Antiseptic Rub TFM, 81 FR 42912 (2016, June).

22. Personal Care Products Council, American Cleaning Institute. (2014). *Safety and Effectiveness of Consumer Antiseptics; Topical Antimicrobial Drug Products for over-the-Counter Human Use; Proposed Amendment of Final Monograph; Reopening of Administrative Record.* Retrieved from http://www.cleaninginstitute.org/assets/1/Page/ACI-PCPC-comments-to-FDA-06162014.pdf

23. Safety and Effectiveness of Consumer Antiseptics; Topical Antimicrobial Drug Products for Over-the-Counter Human Use, 21 CFR 310 (2016). Retrieved from https://www.federalregister.gov/documents/2016/09/06/2016-21337/safety-and-effectiveness-of-consumer-antiseptics-topical-antimicrobial-drug-products-for

24. Zhao, P., Zhao, T., Doyle, M. P., Runbino, J. R., & Mang, J. (1998). Development of a model for evaluation of microbial cross-contamination in the kitchen. *Journal of Food Protection, 61,* 960–963.

25. Chen, Y., Jackson, K. M., Chea, F. P., & Schaffer, D. W. (2001). Quantification and variability analysis of bacterial cross-contamination rates in common food service tasks. *Journal of Food Protection, 67*, 72–80.

26. Minnesota Department of Health (MDH). (n.d.). *Where do germs hide?* [A flyer explaining that the bulk of germs are hiding where you least suspect]. Retrieved from http://www.health.state.mn.us/handhygiene/why/hide.html

27. De Wit, J. C., & Kampelmacher, E. H. (1981). Some aspects of microbial contamination of hands of workers in food industries. *Zentralblatt fur Bakteriologie Mikrobiologie and Hygiene, 172*, 290–400.

28. World Health Organization (WHO). (2001). *WHO Surveillance Program for Control of Foodborne Infections and Intoxications in Europe* (Seventh Report, 1993–1998). Berlin, Germany.

29. Henry Ford Health System (n.d.). *Yuck Factor May Boost Hand Hygiene Compliance*]. Retrieved from http://www.henryford.com/news/2016/06/yuck-factor -may-boost-hand-hygiene-compliance

30. Colorado Historic Newspapers Collection (CHNC). (1900, June 14). Bacteria in stair railing. *Daily Journal* (Telluride). Retrieved from https:// www.coloradohistoricnewspapers.org/cgi-bin/colorado?a=d&d=DJT1900 0614.2.18#

31. Yeh, P. J., Simon, D. M., Millar, J. A., Alexander, H. F., & Franklin, D. (2011). A diversity of antibiotic-resistant *Staphylococcus* spp. in a public transportation system. *Osong Public Health and Research Perspectives, 2*(3), 202–209; Lutz, J. K., van Balen, J., Crawford, J. M., Lee, J., Nava-Hoet, R. C., & Hoet, A. E. (2014). Methicillin-resistant *Staphylococcus aureus* in public transportation vehicles (buses): Another piece to the epidemiologic puzzle. *American Journal of Infection Control, 42*(12), 1285–1290.

32. Soge, O. O., Meschke, J. S., No, D. B., & Roberts, M. C. (2009). Characterization of methicillin-resistant *Staphylococcus aureus* and methicillin-resistant coagulase-negative *Staphylococcus* spp. isolated from U.S. West Coast public marine beaches. *Journal of Antimicrobial Chemotherapy, 64*(6), 1148–1155.

CHAPTER 6: HAND DRYERS, OR BACTERIA BLOWERS?

1. Centers for Disease Control and Prevention (CDC). (2010). Handwashing: Clean hands save lives. Retrieved from http://www.cdc.gov/handwashing/;

CDC. (2002). *Guideline for hand hygiene in health-care settings. Morbidity and Mortality Weekly Report, 51*(RR-16). Retrieved from https://www.cdc.gov/mmwr/PDF/rr/rr5116.pdf; U.S. Food and Drug Association (FDA). (2017). *Retail food protection: Employee health and personal hygiene handbook* (updated 2017). Retrieved from https://www.fda.gov/Food/GuidanceRegulation/Retail FoodProtection/IndustryandRegulatoryAssistanceandTrainingResources/ucm113827.htm

2. Patrick, D. R., Findon, G., & Miller, T. E. (1997). Residual moisture determines the level of touch-contact-associated bacterial transfer following hand washing. *Epidemiology and Infection, 119*, 319–325. (Figure 1 in this chapter was adapted from this publication.)

3. Collins, F., & Hampton, S. (2005). Hand-washing and methicillin-resistant *Staphylococcus aureus. British Journal of Nursing, 14*(13), 703–707.

4. Joseph, T., Baah, K., Jahanfar, A., & Dubey, B. (2015). A comparative life cycle assessment of conventional hand dryer and roll paper towel as hand drying methods. *Science of the Total Environment, 515–516*, 109–117.

5. Kimmitt, P., & Redway, K. (2016). Evaluation of the potential for virus dispersal during hand drying: A comparison of three methods. *Journal of Applied Microbiology, 120*(2), 478–486.

6. Redway, K., Knights, B., & Johnson, K. (1995). A comparison of the anti-bacterial performance of two hand drying methods (textile towels and warm air dryers) after artificial contamination and washing of hands. A study by the Applied Ecology Research Group, University of Westminster (London, UK) for ELIS.

7. Knights, B., Evans, C., Barrass, S., & McHardy, B. (1993). Hand drying: An assessment of efficiency and hygiene of different methods. A survey carried out by the Applied Ecology Research Group for the Association of Makers of Soft Tissue Papers: University of Westminster (London, UK).

8. Redway, K., Knights, B., Bozoky, Z., Theobald, A., & Hardcastle, S. (1994). Hand drying: A study of bacterial types associated with different hand drying methods and with hot air dryers. A study by the Applied Ecology Research Group, University of Westminster (London, UK) for the Association of Makers of Soft Tissue Papers.

9. Snelling, A. M., Saville, T., Stevens, D., & Beggs, C. B. (2010). Comparative evaluation of the hygienic efficacy of an ultra-rapid hand dryer vs. conventional warm air hand dryers. *Journal of Applied Microbiology, 110*, 19–26.

10. Best, E. L., & Redway, K. (2015). Comparison of different hand-drying methods: The potential for airborne microbe dispersal and contamination. *Journal of Hospital Infection, 89,* 215–217.

11. Dawson, P., Northcutt, J., Parisi, M., & Han, I. (2016). Bioaerosol formation and bacterial transfer from commercial hand dryers. *Journal of Food: Microbiology, Safety and Hygiene, 1,* 108.

12. Excel Dryer. (2011). Technical data sheet for Xlerator hand dryer. Retrieved from https://www.exceldryer.com/pdfs/XleratorMasterSpec.pdf

13. Statistical Analysis System (SAS). (2015). SAS OnDemand for Academics. Retrieved from https://www.sas.com/en_us/software/on-demand-for-academics.html

14. See note 13.

15. Yamamoto, Y., Kazuhiro, U., & Takahashi, Y. (2005). Efficiency of hand drying for removing bacteria from washed hands: Comparison of paper towel drying with warm air drying. *Infection Control in Hospital Epidemiology, 26,* 316–320.

16. Best, E. L., Parnell, M. H., & Wilcox, M. H. (2014). Microbiological comparison of hand-drying methods: The potential for contamination of the environment, user and bystander. *Journal of Hospital Infection, 88,* 199–206; Knights, B., Redway, K., & Edwards, V. (1997). Hand washing habits in public toilets and the bacterial contamination of the hands before and after washing. A study by the Applied Ecology Research Group, University of Westminster (London, UK) for DEB; Taylor, J. H., Brown, K., Toivenen, J., & Holah, J. T. (2000). A microbial evaluation of warm air hand driers with respect to hand hygiene and the washroom environment. *Journal of Applied Microbiology, 89,* 910–919.

17. World Union of Wholesale Markets (WUWM). (2009). Community guide to good hygienic practices specific to wholesale market management in the European Union. Retrieved from http://www.europeantissue.com/wp-content/uploads/HACCP-and-hygiene.wholesale-market.WUWM-Nov-2009.pdf

18. See note 2.

19. See note 9.

20. Merry, A. F., Miller, T. E., Findon, G., Webster, C. S., & Neff, S. P. (2001). Touch contamination levels during anesthetic procedures and their relationship to hand hygiene procedures: A clinical audit. *British Journal of Anesthesia, 87,* 291–294.

21. Huang, C., Ma, W., & Stack, S. (2012). The hygienic efficacy of different hand-drying methods: A review of the evidence. *Mayo Clinic, 87*(8), 791–798.

22. Mendes, M. F., & Lynch, D. J. (1976). A bacteriological survey of washrooms and toilets. *Journal of Hygiene, 76,* 183–190; Scott, E., & Bloomfield, S. F. (1985). A bacteriological investigation of the effectiveness of cleaning and disinfection procedures for toilet hygiene. *Journal of Applied Bacteriology, 59,* 291–297.

23. Ngeow, Y. F., Ong, H. W., & Tan, P. 1989. Dispersal of bacteria by an electric air hand dryer. *Malaysian Journal of Pathology, 11,* 53–56; Redway, K., & Fawdar, S. (2008). A comparative study of three different hand drying methods: Paper towel, warm air dryer, jet air dryer. Commissioned by the European Tissue Symposium (ETS). Retrieved from http://www.europeantissue.com/pdfs/090402 -2008%20WUS%20Westminster%20University%20hygiene%20study,%20 nov2008.pdf

24. European Tissue Symposium (ETS). (2009). Are air dryers hygienic? Westminster University Study shows that electric hand dryers in public toilets increase bacteria transmission risk. Retrieved from http:// europeantissue.com/pdfs/090618%20Westminster%20University%20 Study%20key%20findings.pdf; Convenience Store Decisions Staff. (2009, June). Paper towels prove most hygienic in restroom study. Retrieved from https://www.cstoredecisions.com/2009/06/08/paper-towels -prove-most-hygienic-in-restroom-study/; Huang, C., Ma, W. J., & Stack, S. (2012).The hygienic efficacy of different hand-drying methods: A review of the evidence. *Mayo Clinic Proceedings, 87*(8), 791–798. Retrieved from https:// www.researchgate.net/publication/225096470_The_Hygienic_Efficacy_of_ Different_Hand-Drying_Methods_A_Review_of_the_Evidence

25. See note 4.

26. Gustafson, D. R., Vetter, E. A., Larson, D. R., Ilstrup, D. M., Maker, M. D., Thompson, R. L., & Cockerill, F. R. (2000). Effects of 4 hand-drying methods on removing bacteria from washed hands: A randomized trial. *Mayo Clinic Proceedings, 75,* 705–708. Retrieved from http://www.americandryer.com/pdf/ MayoClinicHygieneReport.pdf

PART 3: TRANSPORT MECHANISMS

1. Alton Brown (Writer/Director/Host). (2002). Dip madness [Television series episode]. In *Good Eats,* Season 6, Episode 9, on the Food Network.

2. DeNoon, D. J. (2005). "Pox parties" pooh-poohed. *WebMD.* Retrieved from

https://www.webmd.com/children/vaccines/news/20050929/pox-parties
-pooh-poohed#1

3. Zamosky, L. (n.d.). Is dirt good for kids? Are parents keeping things too clean for their kids' good? *WebMD*. Retrieved from http://www.webmd.com/parenting/features/kids-and-dirt-germs#1

4. Severson, K. M., Mallozzi, M., Dirks, A., & Knight, K. L. (2010). B cell development in GALT: Role of bacterial superantigen-like molecules. *Journal of Immunology*, *184*(12), 6782–6789.

5. Padgett, D. A., & Glaser, R. (2003). How stress influences the immune response. *Trends in Immunology*, *24*(8) 444–448.

6. Kolenbrander, P. E., Andersen, R. N., Blehert, D. S., Egland, P. G., Foster, J. S., & Palmer, R. J., Jr. (2002). Communication among oral bacteria. *Microbiology and Molecular Biology Reviews*, *66*, 486–505.

7. Sumi, Y., Miura, H., Michiwaki, Y., Nagaosa, S., & Nagaya, M. (2006). Colonization of dental plaque by respiratory pathogens in dependent elderly. *Archives of Gerontology and Geriatrics*, *44*(2), 119–124.

8. James, A., Pitchford, J. W., & Plank, M. J. (2007). An event-based model of superspreading in epidemics. *Proceedings of the Royal Society*, *274*, 741–747.

9. World Health Organization (WHO). (2006). Almost a quarter of all disease caused by environmental exposure. Retrieved from http://www.who.int/mediacentre/news/releases/2006/pr32/en/

10. Boone, S. A., & Gerba, C. P. (2007). Significance of fomites in the spread of respiratory and enteric viral disease. *Applied and Environmental Microbiology*, *73*(6), 1687–1696; Sattar, S. A., Springthorpe, S., Manu, S., Gallant, M., Nair, R. C., Scott, E., & Kin, J. (2001). Transfer of bacteria from fabric to hands and other fabrics: Development and application of a quantitative model using *Staphylococcus aureus* as a model. *Journal of Applied Microbiology*, *90*(6), 962–970.

11. Pittet, D., Allegranzi, B., Sax, H., Dharan, S., Pessoa-Silva, C. L., Donaldson, L., & Boyce, J. M. (2006). Evidence-based model for hand transmission during patient care and the role of improved practices. *Lancet Infectious Diseases*, *6*(10), 641–652; Marriot, N., & Gravari, B. (2006). *Principles of Food Sanitation*. 5th ed., Food Science Text Series. New York, NY: Springer (p. 83); Rocourt, J. B., & Cossart, P. (1997). *Listeria monocytogenes*. In M. P. Doyle, L. R. Beuchat, & T. J. Montville (Eds.), *Food Microbiology—Fundamentals and Frontiers* (pp. 337–352). Washington, DC: ASM Press; Rose, J. B., & Slifko, T. R. (1999). *Giar-*

dia, Cryptosporidium, and *Cyclospora* and their impact on foods: A review. *Journal of Food Protection, 62,* 1059–1070.

CHAPTER 7: THINGS YOU PUT IN YOUR DRINK

1. Bailey, S. (2014, November). A guide to the non-alcoholic beverage industry. *Market Realist.* Retrieved from http://marketrealist.com/2014/11/guide-non -alcoholic-beverage-industry/; PR Newswire. (2017). Alcoholic beverage market expected to reach $1,594 billion, globally, by 2022—Allied Market Research. Retrieved from http://www.prnewswire.co.in/news-releases/alcoholic-bev erages-market-expected-to-reach-1594-billion-globally-by-2022---allied -market-research-618354503.html

2. Party Drink Calculator. (2017). Where your party planning begins. Retrieved from http://www.thatsawrapper.com/PT-DrinkCalculator.htm

3. Jumma, P. A. (2000). Hand hygiene. Simple and complex. *International Journal of Infectious Diseases, 9,* 3–14.

4. Falcao, J. P., Dias, A. M. G., Correa, E. F., & Falcao, D. P. (2002). Microbiological quality of ice used to refrigerate foods. *Food Microbiology, 19,* 269–276; Gerokomou, V., Voidarou, C., Vatopoulos, A., Velonakis, E., Rozos, G., Alexopoulos, A., . . . Akrida-Demertizi, K. (2011). Physical, chemical and microbiological quality of ice used to cool drinks and foods in Greece and its public health implications. *Anaerobe, 17*(6), 351–353.

5. Levine, W. C., Stephenson, W. T., & Craun, G. F. (1990). Waterborne disease outbreaks, 1986–1988. Centers for Disease Control (CDC). Retrieved from https://www.cdc.gov/mmwr/preview/mmwrhtml/00001596.htm

6. Ries, A. A., Vugia, D. J., Beingolea, L., Palacios, A. M., Vasquez, E., Wells, J. G., . . . Tauxe, R. V. (1992). Cholera in Piura, Peru: A modern urban epidemic. *Journal of Infectious Diseases, 166*(6), 1429–1433; Wilson, I. G., Hogg, G. M., & Barr, J. G. (1997). Microbiological quality of ice in hospital and community. *Journal of Hospital Infection, 36*(3), 171–180.

7. Falcao, J. P., Falcao, D. P., & Gomez, T. A. T. (2004). Ice as a vehicle for diarrheagenic *Escherichia coli. International Journal of Food Microbiology, 91*(1), 99–103.

8. Lateef, A., Oloke, J. K., Gueguim Kana, E. B., & Pacheco, E. (2006). The microbi-

ological quality of ice used to cool drinks and foods in Ogbomoso Metropolis, Southwest, Nigeria. *Internet Journal of Food Safety, 8*, 39–43.

9. Stout, J. E., Yu, V. L., & Muraca, P. (1985). Isolation of *Legionella pneumophila* from the cold water of hospital ice machines: Implications for origin and transmission of the organism. *Infection Control and Hospital Epidemiology, 6*(4), 141–146; Panwalker, A. P., & Fuhse, A. L. (1986). Nosocomial *Mycobacterium gordonae* pseudoinfection from contaminated ice machines. *Infection Control and Hospital Epidemiology, 7*(2), 67–70; Lassucq, S., Baltch, A. L., Smith, R. P., Smithwick, R. W., Davis, B. J., Desjardin, E. K., . . . Cohen, M. L. (1988). Nosocomial *Mycobacterium fortuitum* colonization from a contaminated ice machine. *American Review of Respiratory Disease, 138*(4), 891–894; Wilson, I. G., Hogg, G. M., & Barr, J. G. (1997). Microbiological quality of ice in hospital and community. *Journal of Hospital Infection, 36*(3), 171–180.

10. Nichols, G., Gillespie, I., & de Louvois, J. (2000). The microbiological quality of ice used to cool drinks and ready-to-eat food from retail and catering premises in the United Kingdom. *Journal of Food Protection, 63*(1), 78–82.

11. Faecal bacteria "in ice in Costa, Starbucks and Caffe Nero." (2017, June). *BBC News.* Retrieved from http://www.bbc.com/news/business-40426228

12. Adegoke, G. O., Iwahashi, H., Komatsu, Y., Obuchi, K., & Iwahashi, Y. (2000). Inhibition of food spoilage yeasts and aflatoxigenic moulds by monoterpenes of the spice *Aframonum danielli. African Journal of Biotechnology, 2*(9), 254–263; Adeleye, I. A., & Opiah, L. (2003). Antimicrobial activity of extracts of local cough mixtures on upper respiratory tract bacterial pathogens. *West Indian Medical Journal, 52*(3), 188–190; Belletti, N., Ndagijimana, M., Sisto, C., Guerzoni, M. E., Lanciotti, R., & Gardini, F. (2004). Evaluation of the antimicrobial activity of citrus essences on *Saccharomyces cerevisiae. Journal of Agricultural and Food Chemistry, 52*(23), 6932–6938; Caccioni, D. R., Guizzardi, M., Biondi, D. M., Renda, S., & Ruberto, G. (1998). Relationship between volatile components of citrus fruit essential oils and antimicrobial action on *Penicillium digitatum* and *Penicillium italicum. International Journal of Food Microbiology, 43*(1–2), 73–79; Dada, J. D., Alade, P. I., Ahmad, A. A., & Yadock, L. H. (2002). Antimicrobial activities of some medicinal plants from Soba-Zaria, Nigeria. *Nigerian Quarterly Journal of Hospital Medicine, 12*(1–4), 55–56; Saleem, M., Afza, N., Anwar, M. A., Hai, S. M., & Ali, M. S. (2003). A comparative study of

essential oils of *Cymbopogon citratus* and some members of the genus *Citrus*. *Natural Product Research, 17*(5), 369–373.

13. Fletcher, P., Harman, S., Boothe, A., Doncel, G., & Shattock, R. (2008). Preclinical evaluation of lime juice as a topical microbicide candidate. *Retrovirology, 5*, 3–13. Retrieved from https://retrovirology.biomedcentral.com/track/pdf/10.1186/1742-4690-5-3?site=retrovirology.biomedcentral.com

14. Gardner, J. M., Feldman, A. W., & Zablotowicz, R. M. (1982). Identity and behavior of xylem-residing bacteria in rough lemon roots of Florida citrus trees. *Applied and Environmental Microbiology, 43*(6), 1335–1342.

15. Rusin, P., Maxwell, S., & Gerba, C. (2002). Comparative surface-to-hand and fingertip-to-mouth transfer efficiency of gram-positive bacteria, gram-negative bacteria, and phage. *Journal of Applied Microbiology, 93*(4), 585–592.

16. Martinez-Gonzalez, N. E., Hernandez-Herrera, A., Martinez-Chavez, L., Rodriguez-Garcia, M. O., Torres-Vitela, M. R., Mota de la Garza, L., & Castillo, A. (2003). Spread of bacterial pathogens during preparation of freshly squeezed orange juice. *Journal of Food Protection, 66*(8), 1490–1494.

17. Fendler, E. J., Dolan, M. J., & Williams, R. A. (1998). Hand washing and gloving for food protection. Part I. Examination of the evidence. *Dairy and Food Environmental Sanitation, 18*, 814–823.

18. Montville, R., Chen, Y., & Schaffner, D. W. (2002). Risk assessment of hand washing efficacy using literature and experimental data. *International Journal of Food Microbiology, 83*, 305–313.

19. Rocourt, J. B., & Cossart, P. (1997). Listeria monocytogenes. In M. P. Doyle, L. R. Beuchat, & T. J. Montville (Eds.), *Food microbiology–Fundamentals and frontiers* (pp. 337–352). Washington, DC: ASM Press; Rose, J. B., & Slifko, T. R. (1999). *Giardia, Cryptosporidium* and *Cyclospora* and their impact on foods: A review. *Journal of Food Protection, 62*, 1059–1070.

20. Kusumaningrum, H. D., Riboldi, G., Hazelberger, W. C., & Beumer, R. R. (2003). Survival of foodborne pathogens on stainless steel surfaces and cross-contamination to foods. *International Journal of Food Microbiology, 85*, 227–236; Moore, C. M., Sheldon, B. W., & Jaykus, L-E. (2003). Transfer of *Salmonella* and *Campylobacter* from stainless steel to romaine lettuce. *Journal of Food Protection, 66*(12), 2231–2236; Rodriguez, A., & McLandsborough, L. A. (2007). Evaluation of the transfer of *Listeria* monocytogenes from stainless steel and polyethylene to bologna and American cheese. *Journal of Food Protection*,

70(3), 600–606; Keskinen, L. A., Todd, E. D., & Ryser, E. T. (2008). Transfer of surface-dried *Listeria* monocytogenes from stainless steel knife blades to roast turkey breast. *Journal of Food Protection, 71*(1), 176–181; Marples, R. R., & Towers, A. G. (1979). A laboratory model for the investigation of contact transfer of micro-organisms. *Journal of Hygiene, 82*(2), 237–248; Sattar, S. A., Springthorpe, S., Mani, S., Gallant, M., Nair, R. C., Scott, E., & Kain, J. (2001). Transfer of bacteria from fabrics to hands and other fabrics: Development and application of a quantitative method using *Staphylococcus aureus* as a model. *Journal of Applied Microbiology, 90*, 962–970; Scott, E., & Bloom-field, S. F. (1990). The survival and transfer of microbial contamination via cloth, hands and utensils. *Journal of Applied Microbiology, 68*(3), 271–278; Legg, S. J., Khela, N., Madie, P., Fenwick, S. G., Quynh, V., & Hedderly, D. I. (1999). A comparison of bacterial adherence to bare hands and gloves following simu-lated contamination from a beef carcass. *International Journal of Food Micro-biology, 53*(1), 69–74; Heal, J. S., Blom, A. W., Titcomb, D., Taylor, A., Bowker, K., & Hardy, J. R. W. (2003). Bacterial contamination of surgical gloves by water droplets spilt after scrubbing. *Journal of Hospital Infection, 53*(2), 136–139; Montville, R., & Schaffner, D. W. (2003). Inoculum size influences bacterial cross-contamination between surfaces. *Applied and Environmental Micro-biology, 69*(12), 7188–7193; Blom, A. W., Gozzard, C., Heal, J., Bowker, K., & Estela, C. M. (2002). Bacterial strike-through of re-usable surgical drapes: The effect of different wetting agents. *Journal of Hospital Infection, 52*(1), 52–55; Shale, K., Jacoby, A., & Plaatjies, Z. (2006). The impact of extrinsic sources on selected indicator organisms in a typical deboning room. *Inter-national Journal of Environmental Health Research, 16*(4), 263–272; Merry, A. F., Miller, T. E., Findon, G., Webster, C. S., & Neff, S. P. W. (2001). Touch con-tamination levels during anesthetic procedures and their relationship to hand hygiene procedures: A clinical audit. *British Journal of Anesthesia, 87*, 291–294.

21. Dawson, P., Han, I., Buyukyavuz, A., Aljeddawi, W., Martinez-Dawson, R., Downs, R., . . . Ellis, V. (2017). Transfer of *Escherichia coli* to lemons slices and ice during handling. *Journal of Food Research, 6*(4), 111.

22. Statistical Analysis System (SAS). (2015). SAS OnDemand for Academics. https://www.sas.com/en_us/software/on-demand-for-academics.html

23. See note 22.

24. Patrick, D. R., Findon, G., & Miller, T. E. (1997). Residual moisture determines the level of touch-contact-associated bacterial transfer following hand washing. *Epidemiology and Infection, 119*(3), 319–325; Perez-Rodriguez, F., Valero, A., Carrasco, E., Garcia, R. M., & Zurera, G. (2008). Understanding and modelling bacterial transfer to foods: A review. *Trends in Food Science and Technology, 19*(3), 131–144.

25. Gill, C. O., & Jones, T. (2002). Effects of wearing knitted or rubber gloves on the transfer of *Escherichia coli* between hands and meat. *Journal of Food Protection, 65*(6), 1045–1048.

26. Escartin, E. F., Ayala, A. C., & Lozano, J. S. (1989). Survival and growth of *Salmonella* and *Shigella* on sliced fresh fruit. *Journal of Food Protection, 52*(7), 471–472.

27. See note 20.

28. See note 4.

29. See note 17.

30. See note 20.

31. Hampikyan, H., Bingol, E. B., Cetin, O., & Colak, H. (2017). Microbiological quality of ice and ice machines in food establishments. *Journal of Water and Health, 15*(3), 410–417.

32. See note 4.

33. Chen, Y., Jackson, K. M, Chea, F. P., & Schaffner, D. W. (2001). Quantification and variability analysis of bacterial cross-contamination rates in common food service tasks. *Journal of Food Protection, 64*(1), 72–80.

34. Bloomfield, S. F., & Scott, E. (1997). Cross-contamination and infection in the domestic environment and the role of chemical disinfectants. *Journal of Applied Microbiology, 83,* 1–9.

35. Ibrahim, Y. K., & Ogunmodede, M. S. (1991). Growth and survival of *Pseudomonas aeruginosa* in some aromatic waters. *Pharmaceutica acta Helvetiae, 66*(9–10), 286–288.

36. Dickens, D. L., DuPont, H. L., & Johnson, P. C. (1985). Survival of bacterial enteropathogens in the ice of popular drinks. *Journal of the American Medical Association, 253*(21), 3141–3143; Gaglio, R., Francesca, N., Di Gerlando, R., Mahony, J., De Martino, S., Stucchi, C., Moschetti, G., & Settanni, L. (2017). Enteric bacteria of food ice and their survival in alcoholic beverages and soft drinks. *Food Microbiology, 67,* 17–22.

37. Sheth, N. K., Wisniewski, T. R., & Franson, T. R. (1988). Survival of enteric patho-

gens in common beverages: An in vitro study. *American Journal of Gastroenterology, 83*(6), 658–660.

38. Loving, A. L., & Perz, J. (2007). Microbial flora on restaurant beverage lemon slices. *Journal of Environmental Health, 70*(5), 18–22.

39. Lynch, M. F., Tauxe, R. V., & Hedberg, C. W. (2009). The growing burden of foodborne outbreaks due to contaminated fresh produce: Risks and opportunities. *Epidemiology and Infection, 137*(3), 307–315.

40. See note 33.

CHAPTER 8: CAN I HAVE A TASTE OF THAT?

1. Miller, L., Rozin, P., & Fiske, A. (1998). Food sharing and feeding another person suggest intimacy: Two studies of American college students. *European Journal of Social Psychology, 28*, 423–436.

2. aan het Rot, M., Moskowitz, D., Hsu, Z., & Young, S. (2015). Eating a meal is associated with elevations in agreeableness and reductions in dominance and submissiveness. *Physiology & Behavior, 144*, 103–109.

3. Hui, Y. H., & Sherkat, F. (Eds.). (2005). *Handbook of food science, technology and engineering.* New York, NY: Taylor and Francis.

4. Ways infectious disease spreads. (n.d.). *SA Health*, South Australia Government. Retrieved from http://sahealth.sa.gov.au/wps/wcm/connect/public+content/sa+health+internet/health+topics/health+conditions+prevention+and+treatment/infectious+diseases/ways+infectious+diseases+spread.

5. Dawson, P. L., Purohit, C., & Han. I. Y. (2010). *Bacterial transfer during eating: Transfer of bacteria from mouth to different utensils and from utensils to food.* Saarbrücken, Germany: VDM Verlag Dr. Müller Aktiengesellschaft and Co.

6. Statistical Analysis System (SAS). (2015). SAS OnDemand for Academics. Retrieved from https://www.sas.com/en_us/software/on-demand-for-academics.html

7. Perez-Rodriguez, F., Valero, A., Carrasco, E., Garcia, R. M., & Zurera, G. (2008). Understanding and modeling bacterial transfer to foods: A review. *Trends in Food Science and Technology, 19*(3), 131–144.

8. Kolenbrander, P. E., Anderson, R. N., Blehert, D. S., Egland, P. G., Foster, J. S., & Palmer, J. R. (2002). Communication among oral bacteria. *Microbiology and Molecular Biology Reviews, 66*, 486–505.

9. Harrel, S., & Molinari, J. (2004). Aerosols and splatter in dentistry: A brief review of the literature and infection control implications. *Journal of the American Dental Association, 135*, 429–437.

10. Centers for Disease Control and Prevention (CDC). (2009). Cover your cough. Retrieved from http://www.cdc.gov/flu/protect/covercough.htm

11. New York State Department of Health. (2009). This is how germs spread . . . it's sickening. Retrieved from http://www.health.state.ny.us/publications/7110/

12. World Health Organization (WHO). (2001). *Surveillance program for control of foodborne infections and intoxications in Europe, Seventh report 1993–1998.* Berlin: WHO.

CHAPTER 9: PASS THE POPCORN, PLEASE

1. The Popcorn Board. (2018). *Recipes.* Retrieved from http://www.popcorn.org/EncyclopediaPopcornica/WelcometoPopcornica/HistoryofPopcorn/tabid/106/Default.aspx

2. Gustaitis, J. (2001, October). The explosive history of popcorn: From maize to microwaves, America's best-loved snack food has long been a favorite treat. *American History, 32–36.*

3. Novozymes Bioenergy. (2018). *Think Bioenergy.* Retrieved from http://thinkbioenergy.com/did-you-know-there-were-6-different-types-of-corn/

4. The Popcorn Board. (2018). *Facts & Fun: From seed to snack.* Retrieved from http://www.popcorn.org/Facts-Fun/From-Seed-to-Snack

5. Jumaa, P. A. (2005). Hand hygiene: Simple and complex. *International Journal of Infectious Diseases, 9*(1), 3–14.

6. Judah, G., Donachie, P., Cobb, E., Schmidt, W., Holland, M., & Curtis, V. (2010). Dirty Hands: Bacteria of faecal origin on commuters' hands. *Epidemiology and Infection, 138*(3), 409–414.

7. Baker, K. A., Han, I. Y., Johnson, L., Jones, E., Knight, A., MacNaughton, M., . . . Dawson, P. (2015). Bacterial transfer from hands while eating popcorn. *Food and Nutrition Sciences, 6*, 1333–1338. Retrieved from http://www.scirp.org/journal/fns

8. Statistical Analysis System (SAS). (2015). SAS OnDemand for Academics. https://www.sas.com/en_us/software/on-demand-for-academics.html

9. Kramer, A., Schwebke, I., & Kampf, G. (2006). How long do nosocomial patho-

gens persist on inanimate surfaces? *BMC Infectious Diseases, 6*, 130–138; Clark, R. P., & de Calcin-Goff, M. L. (2009). Some aspects of the airborne transmission of infection. *Interface, 6*(6), 767–782.

10. Scott, E., & Bloomfield, S. F. (1990). The survival and transfer of microbial contamination via cloths, hands and utensils. *Journal of Applied Bacteriology, 68*(3), 271–278.

11. Boyce, J. M., & Pittet, D. (2001). Guideline for hand hygiene in health-care settings. Recommendations of the Healthcare Infection Control Practices Advisory Committee and the HICPAC/SHEA/APIC/IDSA Hand Hygiene Task Force. *CDC Morbidity and Mortality Weekly Report, 51*(RR-16), 1–45; Chiller, K., Selkin, B. A., & Murakawa, G. J. (2001). Skin microflora and bacterial infections of the skin. *Journal of Investigative Dermatology Symposium Proceedings, 6*, 170–174; Larson, E. (2001). Hygiene of the skin: When is clean too clean? *Emerging Infectious Diseases 2001, 7*, 225–339; Pittet, D., Allegranzi, B., Sax, H., Dharan, S., Pessoa, C. L., Donaldson, L., & Boyce, J. M. (2006). Evidence-based model for hand transmission during patient care and the role of improved practices. *Lancet Infectious Diseases, 6*, 641–652.

12. Kolenbrander, P. E., Andersen, R. N., Blehert, D. S., Egland, P. G., Foster, J. S., & Palmer, R. J. (2002). Communication among oral bacteria. *Microbiology and Molecular Biology Reviews, 66*, 486–505.

13. Trevino, J., Ballieu, B., Yost, R., Danna, S., Harris, G., Dejonckheere, J., . . . Dawson, P. (2009). Effect of biting before dipping (double-dipping) chips on the bacterial population of the dipping solution. *Journal of Food Safety, 29*(1), 37–48.

14. Centers for Disease Control (CDC). (2014). Influenza (Flu): Cover your cough. Retrieved from http://www.cdc.gov/flu/protect/covercough.htm

15. Harrel, S., & Molinari, J. (2004). Aerosols and splatter in dentistry: A brief review of the literature and infection control implications. *Journal of the American Dental Association, 135*, 429–437; Sumi, Y., Miura, H., Michiwaki, Y., Nagaosa, S., & Nagaya, M. (2006). Colonization of dental plaque by respiratory pathogens in dependent elderly. *Archives of Gerontology and Geriatrics 2006, 44*(2), 119–124.

16. CleanLink. (2014, March). *ABC's 20/20 swabs movie theaters—and what they find ain't pretty.* Retrieved from http://www.cleanlink.com/news/article/ABCs-2020-Swabs-Movie-Theaters-8212-And-What-They-Find-Aint-Pretty--16824

17. CDC. (n.d.). *Flavorings-related lung disease.* National Institute for Occupational Safety and Health (NIOSH). Retrieved from https://www.cdc.gov/niosh/topics/flavorings/exposure.html

CHAPTER 10: DIP CHIPS AND DOUBLE-DIPPING

1. Statistical Analysis System (SAS). (2015). SAS OnDemand for Academics. https://www.sas.com/en_us/software/on-demand-for-academics.html
2. Tharmaraja, N., & Shah, N. P. (2004). Survival of *Lactobacillus paracasei* subsp. *paracasei, Lactobacillus rhamnosus, Bifidobacterium animalis* and *Propionibacterium* in cheese-based dips and the suitability of dips as effective carriers of probiotic bacteria. *International Dairy Journal, 14,* 1055–1066.
3. Severson, K. M., Mallozzi, M., Dirks, A., & Knight, K. L. (2010). B cell development in GALT: Role of bacterial superantigen-like molecules. *Journal of Immunology, 184*(12), 6782–6789.
4. American Gut. (n.d.). Home page. Retrieved from http://americangut.org/
5. Zhou, L., & Foster, J. (2015). Psychobiotics and the gut-brain axis: In the pursuit of happiness. *Neuropsychiatric Disease and Treatment, 11,* 715–723.
6. Mole, B. M. (2013, January). Repoopulation remedy. *The Scientist.* Retrieved from http://www.the-scientist.com/?articles.view/articleNo/33897/title/Repoopulation-Remedy/
7. Askt, J. (2017, June). Athletes' microbiomes differ from nonathletes. *The Scientist.* Retrieved from http://www.the-scientist.com/?articles.view/articleNo/49450/title/Athletes--Microbiomes-Differ-from-Nonathletes/andutm_campaign=NEWSLETTER_TS_The-Scientist-Daily_2016andutm_source=hs_emailandutm_medium=emailandutm_content=53331681and_hsenc=p2ANqtz-_LIdaAV9faLFgMzTeUC-ZDmT-E-mUwkavtsINOg82B_IeliOj30jGNg64SA1uaPmQ92KLdXn_10GnNhf8L3dmFJOA49wand_hsmi=53331681/
8. Rubio, R., Jifre, A., Aymerich, T., Guardia, M. D., & Garriga, M. (2014). Nutritionally enhanced fermented sausages as a vehicle for potential probiotic lactobacilli delivery. *Meat Science, 96*(2), 937–942; Gray, N. (2014). Pooperoni? Spanish researchers produce low sodium and fat probiotic stool sausage. Retrieved from https://www.nutraingredients.com/Article/2014/03/04/Pooperoni-Spanish-researchers-produce-low-sodium-and-fat-probiotic-stool-sausage

9. Kang, J-G., Kim, S. H., & Ahn, T-Y. (2006). Bacterial diversity in the human saliva from different ages. *Journal of Microbiology, 44*(5), 572–576.

10. Fitzgibbon, E. J., Bartzokas, C. A., Martin, M. V., Gibson, M. F., & Graham, R. (1984). The source, frequency, and extent of bacterial contamination of dental water systems. *British Dental Journal, 159*, 98–100; Peters, E., & McGaw, W. T. (1996). Dental water unit water contamination. *Journal of the Canadian Dental Association, 62*, 492–495; Pankhurst, C. L., Wood, R. G., & Johnson, N. W. (1998). Causes and prevention of microbial contamination of dental water. *International Dental Journal, 48*, 359–368.

EPILOGUE: FOOD MICROBES AND SAFETY

1. Scallan, E., Hoekstra, R. M., Angulo, F. J., Tauxe, R. V., Widdowson, M-A., Roy, S. L., . . . Griffin, P. M. (2011). Foodborne illness acquired in the United States—Major pathogens. *Emerging Infectious Diseases, 17*(1), 7–15. Retrieved from https://wwwnc.cdc.gov/eid/article/17/1/P1-1101_article

2. Gould, L. H., Walsh, K., Vierra, A., Herman, K., Williams, I., Hall, A., & Cole, D. (2013). Surveillance for foodborne disease outbreaks—United States, 1998–2008. *CDC Morbidity and Mortality Weekly Report*, Surveillance Summaries *62*(2). Retrieved from http://www.cdc.gov/mmwr/pdf/ss/ss6202.pdf

3. Bottemiller, H. (2012, January). Annual foodborne illnesses cost $77 billion, study finds. *Food Safety News*. Retrieved from http://www.foodsafety news.com/2012/01/foodborne-illness-costs-77-billion-annually-study -finds/#.VmgS7I-cFPa

4. International Food Information Council (IFIC). (2015, September). Food safety: A communicator's guide to improving understanding. *Food Insight*. Retrieved from http://www.foodinsight.org/foodsafetyguide

5. Fight BAC! Partnership for Food Safety Education. (n.d.). Retrieved from https:// www.fightbac.org

6. U.S. Food and Drug Administration (FDA). About the Center for Food Safety and Applied Nutrition. Retrieved from https://www.fda.gov/AboutFDA/Centers Offices/OfficeofFoods/CFSAN/

7. Centers for Disease Control and Prevention (CDC). (n.d.). *Food Safety*. Retrieved from http://www.cdc.gov/foodsafety/

8. CDC. (n.d.). Water, sanitation, and environmentally-related hygiene: Keeping hands clean. Retrieved from http://www.cdc.gov/healthywater/hygiene/hand/handwashing.html

9. McSwane, D., Rue, N., & Linton, R. (2000). *Essentials of food safety and sanitation.* 2nd ed. Upper Saddle River, NJ: Prentice-Hall (pp. 250–259).

10. U.S. Department of Agriculture (USDA). (2011, May). Cooking meat? Check the new recommended temperatures [Blog post by Diane Van, Manager, USDA Meat and Poultry Hotline]. Retrieved from http://blogs.usda.gov/2011/05/25/cooking-meat-check-the-new-recommended-temperatures/

Index

- - - - - - - - - - -

Note: Page numbers in italics indicate figures.

alkaliphiles, *16*

allergens, undeclared presence of, 28

American Gut project, 224

amino acids, 13, 14

ammonia, 13

amylase, 41

anaerobic bacteria, 18

animalcules, 7

antibiotics, resistance to, 40

antimicrobials, 15, 78

antimicrobial soap, 130

armrests, 210

Aspergillus flavus, 30

Aston University, 50

Australian League, 68

autotrophic bacteria, 13

bacilli, 8–9, *9*, 31, 61–62

Bacillus, *9*, 61–62

   Bacillus cereus, 31

bacteria, 1, 6, 7. *See also* bacterial cells;
   bacterial growth; bacterial transfer; *specific bacteria*

   "aerolized" by hand dryers, 139

   aerosolized, 150–51

   airborne, 103–53

   anaerobic bacteria, 18

   autotrophic, 13

   communication among, 40

   as "communities," 39–40

   environmental modification by,
   223–24

   exposure to, 158–59

   fecal, 41

   found in nearly every ecosystem, 6

   "good," 159, 224, 225

   infectious, 30

   mesophilic, 133

   oral, 160

   planktonic, 39

   probiotic, 224

   recovery methods, 193–95

   resistant, 40

   sessile, 39

   size of, 8

   on surfaces, 37–99 (*see also spe-*
   cific surfaces)

   transport mechanisms, 157–237
   (*see also* bacterial transfer)

   used as food, 13

   used to make fuels, 13

   waterborne, 103–53

bacterial cells, 30

   division of, 21–22, *21*, *23*

   four distinct phases of growth cycle,
   22–24

   life span of, 21

   resistant, 9–10

   shapes of, 8–9, *9*

   size of, 7–8

bacterial growth

   bacterial growth curve, 23–24, *23*

   bacterial growth rate, *24*

   effect of temperature on bacterial
   growth, *20*, 24

bacterial serotypes, 28

bacterial transfer, 165, 166, *170*, 172–
   75, *175*, *185*, *188*, *192*

   dips and, 220–24, 226

   double-dipping and, 220–23

saliva, 195, 197, 210, 225

Salmonella, 20, 24, 26, 27, *27*, 62, 64, 129, 133, 165, 172

  in ice cubes, 174

  on menus, 85–86

  Salmonella enterica, 28, 30, 161

  Salmonella Enteriditis, 63–64, 234–35

  Salmonella Typhimurium, 52–59, *56, 57, 58, 61, 64*, 159, 160–61, 165, 177

  Salmonella Typhi, 176, 223

salsa, 220–23

salt and pepper shakers, 42

sanitation procedures, 4, 83–84

  water temperature and, 232

sanitizers, 105, 119, 126–34, 232

  claims of, 129–30

  resistance to, 40

  studies of, 126–29, *127*

SARS (severe acute respiratory syndrome), 115, 195, 196

sauces, reheating, 235

Savage, Adam, 48

Schanzle, Jessie, 47

Science Channel, 50

Scott, Elizabeth, 209

seafood, raw, 233

seat cushions, 210

Seinfeld, 213

"self-contamination," 209

sell-by date, 231

serotypes, 28

ServSafe protocol, 119–25

ServSafe training program, 104–5, 118–19

Sheldon, Brian, 129

Shigella, 172, 177

  Shigella flexneri, 176

  Shigella sonnei, 176

shopping smart, 231

Shriners Hospital, 98

skin, 97, 117–34. *See also* hand hygiene and sanitation

sneezing, 103, 110, 113, 196, 223

soap, 232. *See also* hand washing

  antibacterial, 131–32

  antifungal, 131

  antimicrobial, 130

  bactericidal, 130

sodium, 14

soups

  reheating, 235

  sharing, 181, 186–89, *188*

Spanish flu pandemic of 1918, 10

speaking, 103, 104

spinach, 29

Spirilla, *9*

spirillum, *9*

spirochete, *9*

spittle, 7

sponges, 232, 233

spores, 30, 31, 61–62

sporulation, 9–10

springwater, 33

stainless steel, 64

staph. *See Staphylococcus*

Staphylococcus, 9, 18, 20, 27, 30, 42, 90–91, 104, 137, 139, 236

  antibiotic resistance and, 97

  on menus, 86, 87–89, 97–98

  staph infections, 181

tuberculosis (TB), 104, 181, 195
turkey, cooking to proper temperatures, 234
TV remotes, 42
20/20, 210
"Typhoid Mary," 160–61, 223

Uecker, Bob, 68
University of Leeds, 148–51
University of Guadalajara, 172
University of Illinois, 47
University of Westminster, 138–39, *138*
urea, 41
urine/sweat (urea), 41
urophagia, 2
U.S. Department of Agriculture (USDA), 25
U.S. Department of Labor, 117–18
use-by date, 231
U.S. Food and Drug Administration (FDA), 25
  antibacterial soaps on, 131–32
  Fight BAC Program, 230
  Food Code, 104–5, 139
  hand-washing and, 104–5
utensils, 62
  sanitizing, 233
  serving food with clean, 235
  sharing, 181, 182–85
  washing, 232, 233

vaccines, 158–59
vacuum packaging, 18
vaginitis, 40

veal, cooking to proper temperatures, 234
vegetables. *See also specific kinds of produce*
  cutting, 233
  washing, 233
Vibrios, *9*
vinyl, 71–72, *73*, 74–77
virons, 11, 12
viruses, 1, 7, 10–13, *12*, 158–59. *See also specific viruses*
  bacteriophage virus, 11
  complex, *12*
  dormant, 11, 12
  foodborne disease and, 10–11
  found in nearly every ecosystem, 11
  helical, *12*
  infectious, 30
  pathogenic enteric, 30
  polyhedral, *12*
  shapes of, 11
  size of, 8
  spherical, *12*
  transmission of, 12
  varieties of, *12*
vitamins, 14
vomiting, 236

water, 157, 158
  microbes and, 103–53
  water activity (a_w), 15, 17–18
  water temperature, 232
whirlpools, 42
Whole Foods Market, 28